# PROBLEM SOLVING IN SOIL MECHANICS

# Problem Solving in Soil Mechanics

A. AYSEN

A.A. BALKEMA PUBLISHERS    LEIDEN / LONDON / NEWYORK / PHILADELPHIA / SINGAPORE

*Library of Congress Cataloging-in-Publication Data*

*Applied for*

TA
710
.A97
2005

.i 12699652

#73514383

Hardback edition: 2003, ISBN 90 5809 531 2
Paperback edition: 2005, ISBN 0 415 38392 7

Also available: Soil Mechanics – Basic Concepts and Engineering Applications by
☐            A. Aysen, ISBN 0 415 38393 5

Cover design: Studio Jan de Boer, Amsterdam, The Netherlands

Published by: A.A. Balkema Publishers, a member of Taylor & Francis Group plc
☐            www.balkema.nl, www.tandf.co.uk, www.crcpress.com

ISBN: 0 415 38392 7
Printed in Great Britain

# Contents

# Preface

*Problem Solving in Soil Mechanics* is primarily designed as a supplement to *Soil Mechanics: Basic Concepts and Engineering Applications*, but can be used as an independent problem solving text, since there is no specific reference to any equation or figure in the main book. This book is written for university students taking first-degree courses in civil engineering, environmental and agricultural engineering. Its main aim is to stimulate problem solving learning as well as facilitating self-teaching. The book is written with the following objectives:

1. To present the solution of unsolved problems of *Soil Mechanics: Basic Concepts and Engineering Applications*.
2. To provide all necessary methods, equations and figures in a clear step by step explanation of the solution to each problem.

Each chapter is composed of three sections: introduction, worked examples and references for further readings. In the introduction section, the main objectives and the range of problems covered are presented. The second section includes those unsolved problems of the corresponding chapter in the main book. The summary of the theory including equations and figures are described within the solution of each problem.

The special structure of the book makes it possible to be used in two, three and four year undergraduate courses in soil mechanics. However, as new and advanced topics are included, the book will also be a valuable resource for the practicing professional engineer.

The use of S.I. units throughout, and frequent references to current international codes of practice and refereed research papers, make the contents universally applicable. This book is written for readers that have prior knowledge in soil mechanics; however, necessary basic information is included in each worked example.

A. Aysen, M.Eng., Ph.D.
aysena@bigpond.com
March 2005

# Acknowledgements

A great debt of gratitude to my teachers Professor J.R.F. Arthur (University of London), late Professor P.W. Rowe (University of Manchester) and late Professor A.W. Bishop (University of London) for teaching me, examining me and qualifying me in the field of Soil Mechanics.

To Professor J.P. Carter (University of Sydney) for his support during my academic life. To Professor S.W. Sloan (University of Newcastle, Australia) for introducing me to the world of numerical analysis in soil mechanics during my years in the university of Newcastle.

To Professor A.S. Balasubramaniam (Griffith University, Australia) for his continuous support. To dear friends Dr A. Kilpatrick and Mr. R. Fulcher for their encouragements and support.

To Dr H. Katebi for his constructive views and guidance. To Mr M. Conway and Mr Y. Rahimi for their valuable teaching and help in soil mechanics laboratory techniques.

I am in debt to my family, especially my wife Pari, for her unwavering support and patience during this project.

A. Aysen, M. Eng., Ph.D.
aysena@bigpond.com
March 2005

CHAPTER 1

# Nature of Soils, Plasticity and Compaction

## 1.1 INTRODUCTION

This chapter encompasses the three major topics in relation to the basic characteristics of soil, and its physical properties. These topics are associated with the mass-volume relationships (Problems 1.1 to 1.5), index properties (Problems 1.7 to 1.9) including particle size distribution (Problems 1.6, 1.11 and 1.12) and compaction (Problem 1.10). The mass-volume relationships describe parameters, which control the engineering behaviour of the soil. These parameters are void ratio $e$, porosity $n$, degree of saturation $S_r$, air content $A_v$, moisture content $m$ or $w$, density of soil $\rho$ (dry, saturated and natural), and density of solids $\rho_s$ or specific gravity of solids $G_s$. The particle (grain) size analysis, and plasticity indices are needed for soil classification. Problems 1.11 and 1.12 show the engineering application of the particle size analysis where the particle size distribution of a mixture is obtained by knowing the particle size distributions of the individual materials within the mixture. Soil compaction is necessary to improve its strength and bearing capacity. In a laboratory technique explained in Problem 1.10 the relationship between moisture content and density is demonstrated to yield the optimum moisture content corresponding to the maximum dry density.

## 1.2 PROBLEMS

Problem 1.1

The following data are given for a specimen of soil:
$M = 221$ g, $M_s = 128$ g, $G_s = 2.7$, $S_r = 75\%$.
Determine the total volume and the porosity of the specimen.
Solution:

Mass related symbols are:
$M_s$ = mass of solids (or dry mass), $M_w$ = mass of water, $M_a$ = mass of air = 0.

$$M = \text{Total mass of soil sample} = M_s + M_w \qquad (1.1)$$

$M = 221.0 = M_s + M_w = 128.0 + M_w$, $M_w = 93.0$ g.

$\rho_w$ = density of water = $1\text{g/cm}^3$ = $1\text{Mg/m}^3$.

$V_w = 93.0 / \rho_w = 93.0 / 1.0 = 93.0\,\text{cm}^3$.

Volume related symbols are defined as follows:
$V_a$ = volume of the air within the voids between particles.
$V_w$ = volume of the water within the voids between particles. $V_s$ = volume of the solids.
$V_v$ = volume of the voids within a given sample $= V_w + V_a$.

$$V = \text{Total volume of soil sample} = V_s + V_v \qquad (1.2)$$

Density of solids $\rho_s$ is the ratio of the mass of the solids to the volume of the solids:

$$\rho_s = \frac{M_s}{V_s} \qquad (1.3)$$

Specific gravity of solids $G_s$ is the ratio of the density of the solids to the density of water:

$$G_s = \frac{\rho_s}{\rho_w} \qquad (1.4)$$

$$\rho_s = M_s / V_s \rightarrow M_s = \rho_s \times V_s = G_s \times \rho_w \times V_s = 2.7 \times 1.0 \times V_s = 128.0,$$
$$V_s = 47.4 \text{ cm}^3.$$

Degree of saturation $S_r$ is the ratio of the volume of the water to the volume of the voids:

$$S_r = \frac{V_w}{V_v} \qquad (1.5)$$

$$S_r = 0.75 = V_w / V_v = 93.0 / V_v \rightarrow V_v = 124.0 \text{ cm}^3, \text{ thus}$$
$$V = V_s + V_v = 47.4 + 124.0 = 171.4 \text{ cm}^3.$$

Void ratio $e$ is the ratio of the volume of the voids to the volume of the solids:

$$e = \frac{V_v}{V_s} \qquad (1.6)$$

Porosity $n$ is the ratio of the volume of voids to the total volume:

$$n = \frac{V_v}{V} \rightarrow n = \frac{V_v}{V_s + V_v} = \frac{e}{1+e} \qquad (1.7)$$

$$n = V_v / V = 124.0 / 171.4 = 0.723 = 72.3\%.$$

Problem 1.2

Dry soil with $G_s = 2.71$ is mixed with 16% by weight of water and compacted to produce a cylindrical sample of 38 mm diameter and 76 mm long with 6% air content. Calculate the mass of the mixed soil that will be required and the void ratio of the sample.

Solution:

$$V = 7.6 \times \frac{\pi (3.8)^2}{4} = 86.19 \text{ cm}^3. \ \rho_s = G_s \rho_w = 2.71 \times 1.0 = 2.71 \text{ g/cm}^3.$$
$$V_s + V_w + V_a = 86.19,$$

$$\frac{M_s}{2.71} + \frac{0.16 M_s}{\rho_w} + 0.06 \times 86.19 = 86.19, \ \rho_w = 1 \text{ g/cm}^3, \text{ hence:}$$

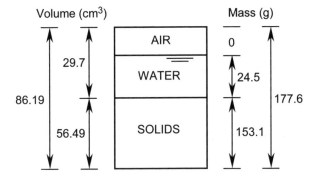

Figure 1.1. Problem 1.2.

$M_s = 153.1 \, \text{g}, \ M = 153.1 + 0.16 \times 153.1 = 177.6 \, \text{g}.$

$M_s = 153.1 = V_s \times \rho_s = V_s \times G_s \times \rho_w = 2.71 V_s,$

$V_s = 153.1/2.71 = 56.49 \, \text{cm}^3.$

$V_v = V - V_s = 86.19 - 56.49 = 29.7 \, \text{cm}^3,$

$e = V_v / V_s = 29.7/56.49 = 0.52.$

The results are shown in Figure 1.1.


Problem 1.3

During a field density test 1850 g of soil was excavated from a hole having a volume of $900 \, \text{cm}^3$. The oven-dried mass of the soil was 1630 g. Determine the moisture content, dry density, void ratio and degree of saturation. $G_s = 2.71$.

Solution:

Moisture content and different types of density are according:

$$w = \frac{M_w}{M_s} \qquad (1.8)$$

$$\rho \, (\text{bulk or wet density}) = \frac{M}{V} \qquad (1.9)$$

$$\rho_d \, (\text{dry density}) = \frac{M_s}{V}, \ \rho_{sat} \, (\text{saturated density}) = \frac{M_s + V_v \rho_w}{V} \qquad (1.10)$$

The relationship between dry density, moisture content and bulk or wet density is:

$$\rho_d = \frac{\rho}{1 + w} \qquad (1.11)$$

$M_w = M - M_s = 1850.0 - 1630.0 = 220.0 \, \text{g}.$

$w = M_w / M_s = 220.0/1630.0 = 0.135 = 13.5\%.$

$\rho_d = M_s / V = 1630.0/900.0 = 1.81 \, \text{g/cm}^3 \text{ or Mg/m}^3.$

$V_s = 1630.0/(2.71 \times 1.0) = 601.48 \, \text{cm}^3$,

$V_v = 900.0 - 601.48 = 298.52 \, \text{cm}^3$.

$e = 298.52/601.48 = 0.496$.

$S_r = V_w/V_v = (220.0/1.0)/298.52 = 0.737 = 73.7\%$.

## Problem 1.4

A soil specimen has a moisture content of 21.4%, void ratio of 0.72, and $G_s = 2.7$. Determine:

(a) bulk density and degree of saturation,

(b) the new bulk density and void ratio if the specimen is compressed undrained until full saturation is obtained.

Solution:

Assume $V = 1 \, \text{m}^3$, thus

$V_v + V_s = 1, e = 0.72 = V_v/V_s$; solving for $V_v$ and $V_s$:

$V_v = 0.419 \, \text{m}^3, V_s = 0.581 \, \text{m}^3$.

(a) $M_s = V_s \times G_s \rho_w = 0.581 \times 2.7 \times 1.0 = 1.569 \, \text{Mg}$.

$M = 1.569 + 1.569 \times 0.214 = 1.905 \, \text{Mg}$.

$\rho = 1.905/1.0 = 1.905 \, \text{Mg/m}^3$.

$S_r = V_w/V_v = (1.569 \times 0.214/1.0)/0.419 = 0.801 = 80.1\%$.

(b) $\rho = M/V = 1.905/(0.581 + 1.569 \times 0.214/1.0) = 2.08 \, \text{Mg/m}^3$.

$e = V_v/V_s = (1.569 \times 0.214/1.0)/0.581 = 0.578$.

## Problem 1.5

The moisture content of a specimen of a clay soil is 22.4%. The specific gravity of the solids is 2.71.

(a) Plot the variation of void ratio with degree of saturation and calculate the void ratio, and the dry and wet densities at 50% saturation,

(b) a sample of this soil with initial degree of saturation of 50% is isotropically compressed to achieve a void ratio of 0.55. Calculate the volume change in terms of percentage of the initial volume. How much of this volume change is due to the outward flow of water from the sample?

Solution:

(a) $S_r = V_w/V_v = (M_s \times 0.224/1.0)/V_v = (V_s \times 2.71 \times 1.0 \times 0.224/1.0)/V_v = 0.607/e$.

The plot of $e$ versus $S_r$ is shown in Figure 1.2.

For $S_r = 50\%$: $e = 0.607 / 0.5 = 1.214$.

Assume $V = 1 \, \text{m}^3$:

$V_v + V_s = 1, e = 1.214 = V_v/V_s$,

$V_v = 0.5483 \, \text{m}^3, V_s = 0.4517 \, \text{m}^3$.

$M_s = V_s G_s \rho_w = 0.4517 \times 2.71 \times 1.0 = 1.224 \, \text{Mg}$.

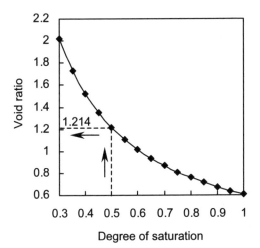

Figure 1.2. Problem 1.5: part (a).

$\rho_d = M_s / V = 1.224 / 1.0 = 1.224 \ \text{Mg/m}^3.$

$M = 1.224 + 1.224 \times 0.224 = 1.50 \ \text{Mg}.$

$\rho = M / V = 1.50 / 1.0 = 1.5 \ \text{Mg/m}^3.$

(b) $e = 0.55 = V_v / 0.4517 \rightarrow V_v = 0.2484 \ \text{m}^3.$

$V = 0.2484 + 0.4517 = 0.7001 \ \text{m}^3.$

Volume change % = $(1.000 - 0.7001) / 1.0 = 0.2999 = 30\%.$

The volume of the outward flow = $(1.224 \times 0.224 / 1.0 - 0.2484) = 0.0258 = 2.6\%$.

The results of part (b) are shown in Figure 1.3.

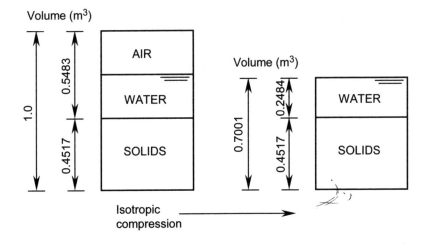

Figure 1.3. Problem 1.5: part (b).

Problem 1.6

The results of a particle size analysis are shown in the table below:

| Sieve size (mm) | Mass retained (g) | Sieve size (mm) | Mass retained (g) |
|---|---|---|---|
| 63 | 0.0 | 4.75 | 50 |
| 37.5 | 26 | 2.36 | 137 |
| 19.0 | 28 | 1.18 | 46 |
| 13.2 | 18 | 0.6 | 31 |
| 9.5 | 20 | 0.212 | 34 |
| 6.7 | 49 | 0.075 | 30 |

The total mass was 469 g. Plot the particle size distribution curve and determine the coefficient of uniformity, coefficient of curvature and soil description.

Solution:

The calculations are summarized in the table below and presented in Figure 1.4, from which:

| Particle size (mm) | Individual mass retained (g) | Individual percentage retained | Cumulative percentage retained | Cumulative percentage finer |
|---|---|---|---|---|
| 63 | 0.0 | 0.00 | 0.00 | 100.00 |
| 37.5 | 26 | 5.54 | 5.54 | 94.46 |
| 19 | 28 | 5.97 | 11.51 | 88.49 |
| 13.2 | 18 | 3.84 | 15.35 | 84.65 |
| 9.5 | 20 | 4.26 | 19.61 | 80.39 |
| 6.7 | 49 | 10.45 | 30.06 | 69.94 |
| 4.75 | 50 | 10.66 | 40.72 | 59.28 |
| 2.36 | 137 | 29.21 | 69.93 | 30.07 |
| 1.18 | 46 | 9.81 | 79.74 | 20.26 |
| 0.6 | 31 | 6.61 | 86.35 | 13.65 |
| 0.212 | 34 | 7.25 | 93.60 | 6.40 |
| 0.075 | 30 | 6.40 | 100.00 | 0.00 |

$D_{10} \approx 0.36$ mm, $D_{30} \approx 2.35$ mm and $D_{60} \approx 4.8$ mm, where $D_{10}$, $D_{30}$, $D_{60}$ are the particle sizes corresponding to 10%, 30% and 60% passing (or finer) respectively. The coefficient of uniformity $C_U$ and the coefficient of curvature $C_C$ are found from:

$$C_U = \frac{D_{60}}{D_{10}} \tag{1.12}$$

$$C_C = \frac{D_{30}^2}{D_{60}D_{10}} \tag{1.13}$$

$$C_U = \frac{D_{60}}{D_{10}} = \frac{4.8}{0.36} = 13.3.$$

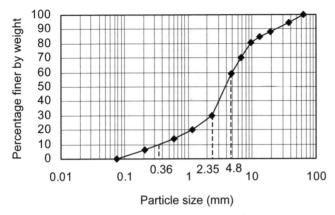

Figure 1.4. Problem 1.6.

$$C_C = \frac{D_{30}^2}{D_{60}D_{10}} = \frac{2.35^2}{4.8 \times 0.36} = 3.2.$$

The soil may be classified as *GW* (well graded gravel).


Problem 1.7

The following data were recorded in a liquid limit test using the Casagrande apparatus. Determine the liquid limit of the soil. Classify the soil assuming plastic limit $PL = 19.8\%$.

| Number of blows | Mass of can (g) | Mass of wet soil + can (g) | Mass of dry soil + can (g) |
|---|---|---|---|
| 8 | 11.80 | 36.05 | 29.18 |
| 16 | 13.20 | 34.15 | 28.60 |
| 27 | 14.10 | 36.95 | 31.16 |
| 40 | 12.09 | 33.29 | 28.11 |

| Number of blows | Moisture content |
|---|---|
| 8 | 0.395 |
| 16 | 0.360 |
| 27 | 0.339 |
| 40 | 0.323 |

The results of the calculations for moisture contents are shown in the table above and Figures 1.5 (moisture content against number of blows) and 1.6 (plasticity chart: ASTM D-2487) from which:

$LL = 34.2\%$.

The soil is classified as *CL*.

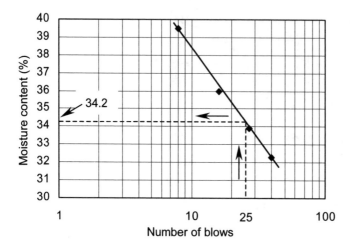

Figure 1.5. Problem 1.7: moisture content against number of blows.

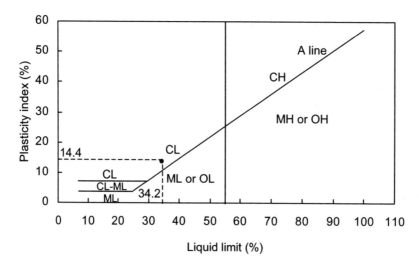

Figure 1.6. Problem 1.7: plasticity chart (ASTM D-2487).

Problem 1.8

The recorded data in a liquid limit test using the cone penetration method are as follows. Determine the liquid limit of the soil.

| Cone penetration (mm) | 14.1 | 18.3 | 22.1 | 27.2 |
|---|---|---|---|---|
| Moisture content (%) | 28.3 | 42.2 | 52.4 | 63.4 |

Solution:
The results are plotted in Figure 1.7 from which $LL = 47\%$.

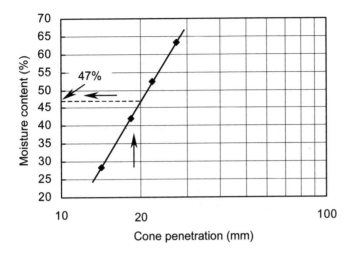

Figure 1.7. Problem 1.8.

Problem 1.9

The maximum and minimum void ratios for a sand are 0.805 and 0.501 respectively. The field density test performed on the same soil has given the following results:

$\rho = 1.81$ Mg/m$^3$, $w = 12.7\%$. Assume $G_s = 2.65$.

Compute the density index.

Solution:

To express the consistency states of sand and gravel, the natural void ratio is compared with the maximum and minimum void ratios obtained in the laboratory. Density index (or relative density) is defined by:

$$I_D = \frac{e_{max} - e}{e_{max} - e_{min}} \qquad \text{(loosest)} \; 0 \le I_D \le 1 \text{(densest)} \qquad (1.14)$$

Dry density in the field is calculated from Equation 1.11:

$$\rho_d = \frac{\rho}{1+w} = \frac{1.81}{1+0.127} = 1.606 \text{ Mg/m}^3.$$

Assume $V = 1$ m$^3$, thus

$M_s = 1.606$ Mg.

$$V_s = \frac{M_s}{G_s \times \rho_w} = \frac{1.606}{2.65 \times 1.0} = 0.606 \text{ m}^3.$$

$V_v = 1.0 - 0.606 = 0.394 \text{ m}^3.$

$e = 0.394 / 0.606 = 0.650$. From Equation 1.14:

$$I_D = \frac{0.805 - 0.650}{0.805 - 0.501} = 0.51.$$

Problem 1.10

The following results are obtained from a standard compaction test:

| Mass of compacted soil (g) | 1920.5 | 2051.5 | 2138.5 | 2147.0 | 2120.0 | 2081.5 |
|---|---|---|---|---|---|---|
| Moisture content (%) | 11.0 | 12.1 | 12.8 | 13.6 | 14.6 | 16.3 |

The specific gravity of the solids is 2.68, and the volume of the compaction mould is 1000 $cm^3$. Plot the compaction curve and obtain the maximum dry density and optimum moisture content. Plot also the 0%, 5% and 10% air void curves. At the maximum dry density, calculate the void ratio, degree of saturation and air content. If the natural moisture content in the field is 11.8%, what will be the possible maximum dry density if the soil is compacted with its natural moisture content?

Solution:

The results of computations are tabulated and shown in Figure 1.8 from which:

$w_{opt} = 13\%$ and $\rho_{dmax} = 1.907$ Mg/m$^3$.

Sample calculations for test number 3:

$\rho = M/V = 2138.5/1000.0 = 2.1385$ g/cm$^3$ or Mg/ m$^3$.

$\rho_d = \rho/(1+w) = 2.1385/(1+0.128) = 1.896$ Mg/m$^3$.

Dry density expressed in terms of $G_s$, $w$, and $A_v$ (air content: $A_v = V_a/V$ ):

$$\rho_d = \frac{G_s \rho_w (1-A_v)}{1+wG_s} \tag{1.15}$$

$$\rho_d = \frac{G_s \rho_w}{1+wG_s} \qquad \text{zero air curve} \tag{1.16}$$

For zero air ($A_v = 0$) with $w = 12.8\%$: $\rho_d = \dfrac{G_s \rho_w}{1+wG_s} = \dfrac{2.68 \times 1.0}{1+0.128 \times 2.68} = 1.995$ Mg/m$^3$.

For 5% air $\rho_d = 1.995 \times 0.95 = 1.896$ Mg/m$^3$.

For 10% air $\rho_d = 1.995 \times 0.9 = 1.796$ Mg/m$^3$.

At the maximum dry density and assuming $V = 1$ m$^3$:

| $w$ (%) | $\rho_d$ (Mg/m$^3$) | $\rho_d$ (Mg/m$^3$): 0 % air | $\rho_d$ (Mg/m$^3$): 5 % air | $\rho_d$ (Mg/m$^3$): 10 % air |
|---|---|---|---|---|
| 11.0 | 1.730 | 2.070 | 1.966 | 1.863 |
| 12.1 | 1.830 | 2.024 | 1.922 | 1.821 |
| 12.8 | 1.896 | 1.995 | 1.896 | 1.796 |
| 13.6 | 1.890 | 1.964 | 1.866 | 1.768 |
| 14.6 | 1.850 | 1.926 | 1.830 | 1.734 |
| 16.3 | 1.790 | 1.865 | 1.772 | 1.679 |
| 17.0 | | 1.841 | 1.749 | 1.657 |
| 18.0 | | 1.808 | 1.717 | 1.627 |

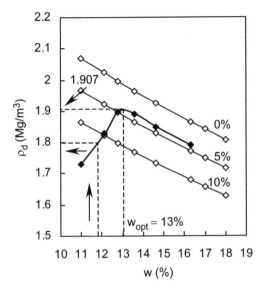

Figure 1.8. Problem 1.10.

$$V_s = \frac{M_s}{G_s \times \rho_w} = \frac{1.907}{2.68 \times 1.0} = 0.711 \text{ m}^3.$$

$$V_v = 1.0 - 0.711 = 0.289 \text{ m}^3.$$

$$e = 0.289/0.711 = 0.406.$$

$$S_r = V_w/V_v = (1.907 \times 0.13/1.0)/0.289 = 0.858 = 85.8\%.$$

$$A_v = V_a/V = (V_v - V_w)/1.0,$$

$$A_v = (0.289 - 1.907 \times 0.13/1.0)/1.0 = 0.041 = 4.1\%.$$

From Figure 1.8 for $w = 11.8\%$ the corresponding dry density is $\approx 1.8 \text{ Mg/m}^3$.

Problem 1.11

The results of two particle size analyses on sand and gravel samples are shown in the first and second columns of the following table. The mass of each specimen is 5 kg. Plot the particle size distribution curve for both specimens. A third specimen is made by mixing two volumes of sand with one volume of gravel. Plot the particle size distribution curve for the mixture assuming that the densities of gravel and sands are equal.

Solution:

The results are tabulated below and shown in Figure 1.9. In the table corresponding to the mixed sample the retained mass on the individual sieve related to the sand is multiplied by 2. Thus the total mass of the assumed mixture is 15 kg. If the particle size distribution for the mixture was given (either as a single curve or a zone bounded by two curves), then a trial and error procedure in terms of the mass ratios of sand and gravel must be adopted until the required distribution is obtained.

| Sieve size (mm) | Individual mass retained (g) | Individual percentage retained | Cumulative percentage retained | Cumulative percentage finer |
|---|---|---|---|---|
| Gravel | | | | |
| 50 | 0 | 0.00 | 0.00 | 100.00 |
| 25 | 375 | 7.50 | 7.50 | 92.50 |
| 12.5 | 825 | 16.50 | 24.00 | 86.00 |
| 6.7 | 1000 | 20.00 | 44.00 | 56.00 |
| 3.45 | 1350 | 27.00 | 71.00 | 29.00 |
| 2 | 1450 | 29.00 | 100.00 | 0.00 |
| Sand | | | | |
| 2 | 0 | 0.00 | 0.00 | 100.00 |
| 1.4 | 500 | 10.00 | 10.00 | 90.00 |
| 1.18 | 200 | 4.00 | 14.00 | 86.00 |
| 0.85 | 550 | 11.00 | 25.00 | 75.00 |
| 0.6 | 1100 | 22.00 | 47.00 | 53.00 |
| 0.425 | 1750 | 35.00 | 82.00 | 18.00 |
| 0.3 | 900 | 18.00 | 100.00 | 0.00 |

In the case where the densities of the given two materials are not equal then the mass of the material 1 on the individual sieve must be multiplied by $m_{r1}$ according:

$$m_{r1} = \frac{V_1}{V_2} \times \frac{\rho_1}{\rho_2} \qquad (1.17)$$

where $V_1 / V_2$ represents the volume ratio (in this example is 2) and $\rho_1$ and $\rho_2$ are the corresponding densities.

| Sieve size (mm) | Individual mass retained (g) | Individual percentage retained | Cumulative percentage retained | Cumulative percentage finer |
|---|---|---|---|---|
| 50 | 0 | 0.00 | 0.00 | 100.00 |
| 25 | 375 | 2.50 | 2.50 | 97.50 |
| 12.5 | 825 | 5.50 | 8.00 | 92.00 |
| 6.7 | 1000 | 6.67 | 14.67 | 85.33 |
| 3.45 | 1350 | 9.00 | 23.67 | 76.33 |
| 2 | 1450 | 9.67 | 33.34 | 66.66 |
| 1.4 | 1000 | 6.67 | 40.01 | 59.99 |
| 1.18 | 400 | 2.66 | 42.67 | 57.33 |
| 0.85 | 1100 | 7.33 | 50.00 | 50.00 |
| 0.6 | 2200 | 14.67 | 64.67 | 35.33 |
| 0.425 | 3500 | 23.33 | 88.00 | 12.00 |
| 0.3 | 1800 | 12.00 | 100.00 | 0.00 |

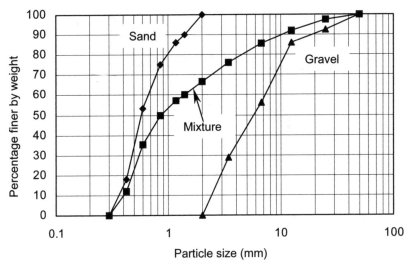

Figure 1.9. Problem 1.11.

Problem 1.12

Re-work Problem 1.11 with the volume ratio (sand to gravel) of 1.0. The densities of gravel and sand are 1.25 Mg/m$^3$ and 1.5 Mg/m$^3$ respectively.

Solution:

The mass of the sand on each individual sieve is multiplied by $m_{r1}$:

$$m_{r1} = \frac{V_1}{V_2} \times \frac{\rho_1}{\rho_2} = 1.0 \times \frac{1.5}{1.25} = 1.2.$$

The corresponding calculations are tabulated and shown in Figure 1.10; total mass of the mixture = 11 kg.

| Sieve size (mm) | Individual mass retained (g) | Individual percentage retained | Cumulative percentage retained | Cumulative percentage finer |
|---|---|---|---|---|
| 50 | 0 | 0.00 | 0.00 | 100.00 |
| 25 | 375 | 3.41 | 3.41 | 96.59 |
| 12.5 | 825 | 7.50 | 10.91 | 89.09 |
| 6.7 | 1000 | 9.10 | 20.01 | 79.99 |
| 3.45 | 1350 | 12.27 | 32.28 | 67.72 |
| 2 | 1450 | 13.18 | 45.46 | 54.54 |
| 1.4 | 600 | 5.45 | 50.91 | 49.09 |
| 1.18 | 240 | 2.18 | 53.09 | 46.91 |
| 0.85 | 660 | 6.00 | 59.09 | 40.91 |
| 0.6 | 1320 | 12.00 | 71.09 | 28.91 |
| 0.425 | 2100 | 19.09 | 90.18 | 9.82 |
| 0.3 | 1080 | 9.82 | 100.00 | 0.00 |

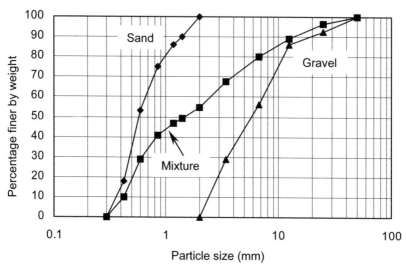

Figure 1.10. Problem 1.12.


## 1.3  REFERENCES AND RECOMMENDED READINGS

ASCE. 1962. Symposium on grouting. *Journal SMFE, ASCE*, 87(SM2): 1-145.

ASTM D-2487. 1998. *Standard classification of soils for engineering purposes (Unified Soil Classification System)*. West Conshohocken, PA: American Society for Testing and Materials.

Atkinson, J. 1993. *An introduction to the mechanics of soils and foundations*. London: McGraw-Hill.

AS.1993. *Geotechnical site investigations*. 3rd edition. NSW, Australia: Standard Association of Australia.

Aysen, A. 2002. *Soil mechanics: Basic concepts and engineering applications*. Lisse: Balkema.

Bowles, J.E. 1996. *Foundation analysis and design*. 5th edition. New York: McGraw-Hill.

BS 5930. 1981. *Code of practice for site investigation*. London: British Standards Institution.

Casagrande, A. 1948. Classifications and identifications of soils. *Translated: ASCE*, 113: 901-991.

Craig, R.F. 1997. *Soil mechanics*. 6th edition. London: E & FN Spon.

Isbell, R.F. 1996. *The Australian soil classification*. Australia, Collingwood: CSIRO Publishing.

Terzaghi, K. & Peck, R. B. 1967. *Soil mechanics in engineering practice*. 2nd edition. New York: John Wiley & Sons.

White, R.E. 1979. *Introduction to the principles and practice of soil science*. Melbourne: Blackwell Scientific.

Yong, R.N, & Warkentin, B.P. 1966. *Introduction to soil behaviour*. New York: The Macmillan Company.

# CHAPTER 2

# Effective Stress and Pore Pressure in Saturated Soils

## 2.1   INTRODUCTION

The stress related to the internal forces acting on the contact points of the particles is termed the *effective stress* whilst the stress within the liquid phase or water is termed *pore pressure*. The combination of these two stresses represents the *total stress* at a point:

$$\sigma = \sigma' + u \tag{2.1}$$

where $\sigma$ is the total normal stress at a point in a specified plane, $\sigma'$ is the effective normal stress on that plane resisted by the particles and $u$ is the pore pressure acting on the plane. In drained loading the effective stress controls the strength of the soil and its deformation. In the undrained conditions the total stress is of major concern. Both conditions may occur in the field in the form of moving from undrained conditions to the drained conditions as the time passes; the process being controlled by the drainage conditions. The change in the effective vertical stress within the soil is the major factor in the consolidation settlement in clay soils. For a soil section composed of $n$ layers, above the point of interest, and each having a thickness of $h_i$, Equation 2.1 yields:

$$\sigma'_v = \sum_{i=1}^{i=n} \gamma h_i - u = \sum_{i=1}^{i=n} \gamma_{ei} h_i \tag{2.2}$$

where $\gamma_{ei}$ is the effective unit weight of each layer. The problems in this chapter are designed to cover three major areas. First, the main definition of the effective stress is investigated through Problems 2.1 and 2.2. Second, the case of the water with a high pressure (artesian conditions) that is located at some depth from the ground surface has been considered (Problems 2.3 and 2.4). The third area is the concept of the increase in the effective vertical stress due to dewatering that is explained in Problem 2.5. Note that only the stresses due to gravity are considered. The cases of external surface loading will be discussed in Chapter 6.

## 2.2   PROBLEMS

Problem 2.1

For the soil profile shown in Figure 2.1 plot the variation of total vertical stress, pore pressure and effective vertical stress with depth.

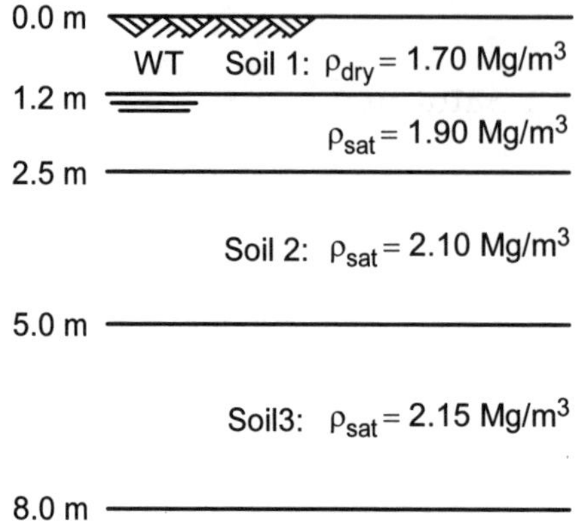

0.0 m
WT
Soil 1: ρdry = 1.70 Mg/m3
1.2 m
ρsat = 1.90 Mg/m3
2.5 m
Soil 2: ρsat = 2.10 Mg/m3
5.0 m
Soil3: ρsat = 2.15 Mg/m3
8.0 m


Figure 2.1. Problem 2.1: soil profile.

Solution:

On the ground surface, $\sigma_v = 0.0$, $u = 0.0$, $\sigma'_v = 0.0$.

At $z = 1.2$ m:

$\sigma_v = 1.7 \times 9.81 \times 1.2 = 20.0$ kPa, $u = 0.0$, $\sigma'_v = 20.0$ kPa.

At $z = 2.5$ m:

$\sigma_v = 20.0 + 1.9 \times 9.81(2.5 - 1.2) = 44.2$ kPa,

$u = 1.0 \times 9.81(2.5 - 1.2) = 12.7$ kPa,

$\sigma'_v = 44.2 - 12.7 = 31.5$ kPa.

At $z = 5.0$ m:

$\sigma_v = 44.2 + 2.1 \times 9.81(5.0 - 2.5) = 95.7$ kPa,

$u = 1.0 \times 9.81(5.0 - 1.2) = 37.3$ kPa,

Stress (kPa)
0
50
100
150
200
Depth (m)
0
1
2
3
4
5
6
7
8
Soil 1
Soil 2
Soil3
Pore pressure
Effective
Total


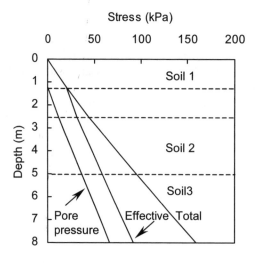

Figure 2.2. Problem 2.1: variation of pore pressure, total and effective vertical stresses with depth.

$\sigma'_v = 95.7 - 37.3 = 58.4\,\text{kPa}.$

At $z = 8.0$ m:

$\sigma_v = 95.7 + 2.15 \times 9.81 \times 3.0 = 159.0\,\text{kPa},$

$u = 1.0 \times 9.81(8.0 - 1.2) = 66.7\,\text{kPa},$

$\sigma'_v = 159.0 - 66.7 = 92.3\,\text{kPa}.$ The results are presented in Figure 2.2.

Problem 2.2

For the given soil profile of Figure 2.3 calculate the effective vertical stress at a depth of 7.5 m.

Solution:

For soil 1 and assuming $V = 1\text{m}^3$:

$S_r = V_w/V_v = 0.5,$

$V_w = M_w/\rho_w = 0.14 \times M_s/1.0,\ V_v = 1 - V_s,\ V_s = M_s/\rho_s = M_s/(G_s \times \rho_w).$

$S_r = \dfrac{V_w}{V_v} = \dfrac{0.14 M_s}{1 - \dfrac{M_s}{2.68 \times 1.0}} = 0.5,\ M_s = 1.531\,\text{Mg}.$

$\rho = (M_s + wM_s)/V = (1.531 + 0.14 \times 1.531)/1.0 = 1.745\,\text{Mg/m}^3.$

$V_s = 1.531/(2.68 \times 1.0) = 0.571\,\text{m}^3,$

$V_v = 1.0 - 0.571 = 0.429\,\text{m}^3,$ thus

$\rho_{sat} = (V_s \times G_s\rho_w + V_v \times \rho_w)/V = (1.531 + 0.429 \times 1.0)/1.0 = 1.960\,\text{Mg/m}^3.$

Soil 2:

Assume $V = 1\,\text{m}^3$, thus

$V_v + V_s = 1, e = V_v/V_s = 5.0;$ solving for $V_v$ and $V_s$:

$V_v = 0.833\,\text{m}^3, V_s = 0.167\,\text{m}^3.$

$\rho_{sat} = (V_s \times G_s\rho_w + V_v \times \rho_w)/V = (0.167 \times 2.0 \times 1.0 + 0.833 \times 1.0)/1.0 = 1.167\,\text{Mg/m}^3.$

Soil 3:

$V_v + V_s = 1, e = V_v/V_s = 1;$ solving for $V_v$ and $V_s$:

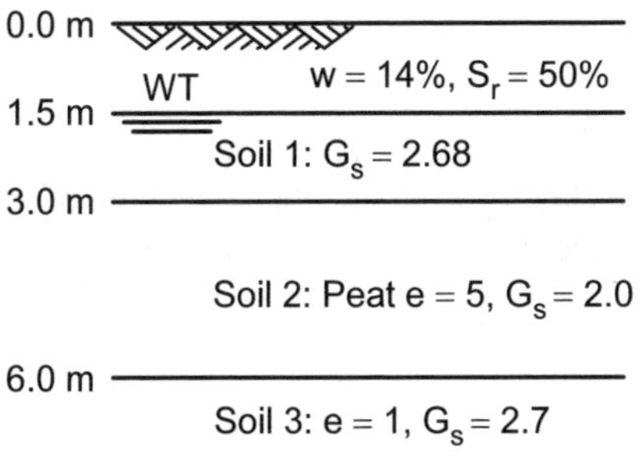

0.0 m

WT     w = 14%, $S_r = 50\%$

1.5 m

Soil 1: $G_s = 2.68$

3.0 m

Soil 2: Peat e = 5, $G_s = 2.0$

6.0 m

Soil 3: e = 1, $G_s = 2.7$

Figure 2.3. Problem 2.2.

$V_v = 0.5\,\text{m}^3, V_s = 0.5\,\text{m}^3.$

$\rho_{sat} = (0.5\times2.7\times1.0+0.5\times1.0)/1.0 = 1.850\,\text{Mg/m}^3.$

At $z = 7.5$ m:

$\sigma_v = (1.745\times1.5+1.960\times1.5+1.167\times3.0+1.850\times1.5)\times9.81,$

$\sigma_v = 116.1\,\text{kPa},$

$u = 1.0\times9.81\times6.0 = 58.9\,\text{kPa},$

$\sigma_v' = 116.1-58.9 = 57.2\,\text{kPa}.$

## Problem 2.3

A clay layer of 4 m thick with $\rho_{sat} = 2$ Mg/m$^3$ is overlain by a 4 m sand with $\rho_{sat} = 1.9$ Mg/m$^3$ and $\rho_{dry} = 1.65$ Mg/m$^3$, the top of this layer being the ground surface. The water table is located 2 m below the ground surface. The clay layer is underlain by a sand stratum that is in artesian conditions with the water level in a standpipe being 4 m above the ground surface.

Calculate the effective vertical stresses at the top and the base of the clay layer. If the dry sand is excavated, in what depth the effective stress at the bottom of the clay layer will become zero?

Solution:

At the top of the clay layer $z = 4$ m:

$\sigma_v = (1.65\times2.0+1.90\times2.0)\times9.81 = 69.6\,\text{kPa},$

$u = 1.0\times9.81\times2.0 = 19.6\,\text{kPa},$

$\sigma_v' = 69.6-19.6 = 50.0\,\text{kPa}.$

At the base of the clay layer $z = 8$ m:

$\sigma_v = 69.6+2.0\times9.81\times4.0 = 148.1\,\text{kPa},$

$u = 1.0\times9.81(8.0+4.0) = 117.7\,\text{kPa},$

$\sigma_v' = 148.1-117.7 = 30.4\,\text{kPa}.$

Assume $D$ is the depth of excavation in the sand layer, thus at the base of the clay layer:

$\sigma_v = 148.1-1.65\times9.81\times D = 148.1-16.2D\,\text{kPa},$

$\sigma_v' = 148.1-16.2D-117.7 = 0,$

$D \approx 1.9\,\text{m} < 2.0$ m (the thickness of the dry sand layer).

## Problem 2.4

A clay layer 10 m thick has a density of 1.75 Mg/m$^3$ and is underlain by sand. The top of the clay is the ground surface. An excavation in the clay layer failed when the depth of the excavation reached to 6.5 m from the ground surface. Calculate the depth of water in a standpipe sunk to the sand layer.

Solution:

Calculate the total vertical stress at the base of the clay:

$\sigma_v = 1.75\times9.81(10.0-6.5) = 60.1\,\text{kPa}.$

The pore pressure on the boundary of sand and clay is:

$u = 1.0 \times 9.81 \times h_w = 9.81 h_w$,

where $h_w$ is the height of water above the boundary.
Set the effective vertical stress at the base of the clay to zero:

$\sigma'_v = 60.1 - 9.81 h_w = 0$,

$h_w = 6.125$ m.

Depth of the piezometric level from the ground surface $= 10.0 - 6.125 = 3.875$ m.


Problem 2.5

A stratum of soil is 15 m thick and its top surface is the ground surface. Formulate the effective vertical stress within the layer if:
(a) the water table is at the ground surface,
(b) the water table is lowered 3 m by pumping.
$\rho_{sat} = 2$ Mg/m$^3$ and $\rho_{dry} = 1.65$ Mg/m$^3$.

Solution:

(a) At a depth $z$ from the ground surface:

$\sigma_v = \rho_{sat} gz = 2.0 \times 9.81 \times z = 19.62z$ kPa,

$u = 1.0 \times 9.81 \times z = 9.81z$ kPa,

$\sigma'_v = 19.62z - 9.981z = 9.81z$ kPa.

(b) At a depth $z > 3.0$ m from the ground surface:

$\sigma_v = \rho_{dry} \times g \times 3.0 + \rho_{sat} g(z - 3.0)$ kPa,

$\sigma_v = 1.65 \times 9.81 \times 3.0 + 2.0 \times 9.81(z - 3.0) = 19.62z - 10.30$ kPa,

$u = 1.0 \times 9.81(z - 3.0) = 9.81z - 29.43$ kPa,

$\sigma'_v = 19.62z - 10.30 - 9.81z + 29.43 = 9.81z + 19.13$ kPa.

The increase in the effective vertical stress after pumping is:

$\Delta\sigma'_v = 9.81z + 19.13 - 9.81z = 19.13$ kPa,

which is independent from the depth of the point of interest.
In general the increase in the effective vertical stress is:

$$\Delta\sigma' = \Delta H \rho_w g - \Delta H g (\rho_{sat} - \rho_m) \tag{2.3}$$

where $\Delta H$ is the magnitude of the drop of the water surface, $\rho_w$ is the density of water and $\rho_m$ is the density of the dewatered zone.
For example for $z > 3$ m:

$\Delta\sigma' = 3.0 \times 1.0 \times 9.81 - 3.0 \times 9.81(2.0 - 1.65) = 19.13$ kPa.

At a depth $z < 3.0$ m from the ground surface:

$\sigma_v = \sigma'_v = \rho_{dry} \times g \times z = 1.65 \times 9.81 \times z = 16.19z$ kPa.

The increase in the effective vertical stress is:

$\Delta\sigma'_v = 16.19z - 9.81z = 6.38z$ kPa.

The formulation at this region is as follows:

$$\Delta\sigma' = gz(\rho_m + \rho_w - \rho_{sat}) \tag{2.4}$$

## 2.3 REFERENCES AND RECOMMENDED READINGS

Aysen, A. 2002. *Soil mechanics: Basic concepts and engineering applications*. Lisse: Balkema.

Bishop, A.W. 1959. The Principal of effective stress. From a lecture in Oslo, Norway 1955. Reprinted in *Teknisk Ukeblad* (39): 859-863.

Klausner, Y. 1991. *Fundamentals of continuum mechanics of soils*. London: Springer-Verlag.

Powrie, W. 1997. *Soil mechanics-concepts and applications*. London: E & FN Spon.

Terzaghi, K. & Peck, R. B. 1967. *Soil mechanics in engineering practice*. 2nd edition. New York: John Wiley & Sons.

CHAPTER 3

# The Movement of Water through Soil

## 3.1 INTRODUCTION

The problems that are solved in this chapter investigate the flow of water through interconnected pores between soil particles in both one and two dimensions. To obtain the *coefficient of permeability k* (or *hydraulic conductivity*) two common laboratory test methods of *constant head* and *falling head* are used. These methods are described through Problems 3.1 to 3.3. The flow is assumed laminar and Darcy's law is valid. Problem 3.4 shows the application of Darcy's law in formulating the flow rate in a two-dimensional flow problem. Problems 3.5 and 3.6 are related to the in-situ test methods of obtaining the coefficient of permeability in unconfined and confined aquifers. The equivalent coefficients of permeability in layered soils (parallel and normal to the stratum) are explained in Problem 3.7. Using a simplified approach the *flow nets* that, describe the seepage flow, are constructed and used to obtain the flow rate under the impermeable dams or sheet piles (Problems 3.8 and 3.9). The flow within a permeable dam is explained by Problems 3.10 and 3.11.

## 3.2 PROBLEMS

Problem 3.1

In a laboratory constant head permeability test, a cylindrical sample 100 mm in diameter and 150 mm high is subjected to an upward flow of 540 ml (cm$^3$)/min. The head loss over the length of the sample is measured to be 360 mm. Calculate the coefficient of permeability in m/s.

Solution:

The hydraulic gradient within the length $L$ is a dimensionless parameter and is defined as the rate of change in total head (or head loss) $\Delta h$ over the length $L$:

$$i = \frac{\Delta h}{L} \qquad (3.1)$$

If we assume the flow obeys Darcy' law:

$$v = ki \qquad (3.2)$$

where $v$ is the velocity and $k$ is the coefficient of permeability of the material. The quantity of water that flows in a unit of time through an area of $A$ or flow rate is:

$$q = \frac{Q}{t} = Av = Aki = Ak\frac{\Delta h}{L} \tag{3.3}$$

Using Equation 3.3 for a constant head permeability test:

$$k = \frac{QL}{\Delta h \times At} = \frac{qL}{\Delta h \times A} \tag{3.4}$$

$$k = \frac{QL}{\Delta h \times At} = \frac{540.0 \times 10^{-6} \times 150.0 \times 10^{-3}}{360.0 \times 10^{-3}(100.0^2 \times \pi/4)10^{-6} \times 1.0 \times 60},$$

$$k = 4.8 \times 10^{-4} \text{ m/s.}$$

Problem 3.2

In a laboratory falling head test, the recorded data are: diameter of the tube = 20 mm, diameter of the cell = 100 mm, length of the sample = 1000 mm. The head measured from the top level of the sample dropped from 800 mm to 600 mm within 1 hour and the temperature of the water was 30 °C. Calculate the coeeficient of permeability at 20 °C. $\eta = 1.005 \times 10^{-3}$ N.s/m$^2$ (at 20 °C), $\eta = 0.801 \times 10^{-3}$ N.s/ m$^2$ (at 30 °C).

Solution:

At the start of the test and at $t = 0$ the head in the vertical capillary tube is equal to $h_1$. The valve on the tube is opened and the time $t$ for the head to fall to $h_2$ is recorded. The coefficient of permeability is calculated from:

$$k = 2.3\frac{aL}{At}\log\frac{h_1}{h_2} \tag{3.5}$$

where $a$ is the internal sectional area of the capillary tube and $A$ is the sectional area of the soil.

$$k = 2.3 \times \frac{20.0^2 \times \pi/4}{100.0^2 \times \pi/4} \times \frac{1000.0 \times 10^{-3}}{1.0 \times 60 \times 60}\log\frac{800.0}{600.0},$$

$$k = 3.19 \times 10^{-6} \text{ m/s.}$$

To include the effect of temperature, the following equation may be used:

$$k_{20} = k_\theta\frac{\eta_\theta}{\eta_{20}} \tag{3.6}$$

where $\theta$ is the temperature of the outflow water in degrees Celsius, $k_{20}$ and $k_\theta$ are the coefficients of permeability at 20 °C and at $\theta$ °C, $\eta_{20}$ and $\eta_\theta$ are the dynamic viscosities of water at 20 °C and $\theta$ °C respectively.

$$k_{20} = 3.19 \times 10^{-6} \times \frac{0.801 \times 10^{-3}}{1.005 \times 10^{-3}},$$

$$k_{20} = 2.54 \times 10^{-6} \text{ m/s.}$$

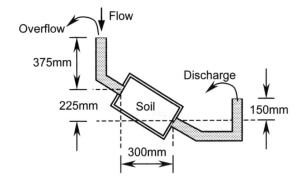

Figure 3.1. Problem 3.3.

Problem 3.3

For the test arrangement shown in Figure 3.1, calculate the volume of water discharged in 20 minutes. The cross-sectional area of the soil is 4000 mm$^2$ and $k = 4.0$ mm/s.

Solution:

$t = 20.0 \times 60 = 1200.0$ s.

$A = 4000.0 \times 10^{-6} = 4.0 \times 10^{-3}$ m$^2$.

$k = 4.0 \times 10^{-3}$ m/s.

$$L = \sqrt{(\frac{300.0}{1000})^2 + (\frac{225.0}{1000})^2} = 0.375 \text{ m}.$$

$$\frac{\Delta h}{L} = \frac{225.0 + 375.0 - 150.0}{375.0} = 1.2.$$

Using Equation 3.3:

$$Q = tAk\frac{\Delta h}{L} = 1200.0 \times 4.0 \times 10^{-3} \times 4.0 \times 10^{-3} \times 1.2,$$

$Q = 23.04 \times 10^{-3}$ m$^3$ = 23.04 l.

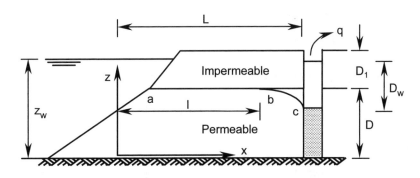

Figure 3.2. Problem 3.4.

Problem 3.4

A long trench is excavated parallel to a river, as shown in Figure 3.2. The soil profile consists of a permeable soil of thickness $D$ confined between two impermeable layers. Initially the water level in the trench is the same as that in the river. Water is pumped out of the trench at a flow rate of $q$.

(a) Formulate $q$ in terms of the geometrical parameters shown in Figure 3.2,

(b) for $L$ (the average horizontal distance between the trench and the river's slope) = 100 m, $D = 5$ m, $z_w = 7$ m, and $k = 4 \times 10^{-5}$ m/s, calculate the flow rate corresponding to a drawdown of $D_w = 2$ m,

(c) calculate $q$ when the water table in the trench is 2 m below the surface of the permeable layer.

Solution:

(a) For full flow between points $a$ and $b$ using Equation 3.3:

$$q = Av = Aki = (D \times 1.0)k\frac{z_w - D}{l},$$

where $l$ is the average horizontal distance between river's slope and the section where the drawdown starts. For the flow between sections $b$ and $c$:

$$q = Av = Aki = (z \times 1.0)k(-\frac{dz}{dx}),$$

$$qdx = -kzdz,$$

$$q\int_{x_b}^{x_c} dx = k\int_{z_c}^{z_b} zdz,$$

$$q(x_c - x_b) = \frac{k}{2}(z_b^2 - z_c^2) \rightarrow q(L - l) = \frac{k}{2}\left[D^2 - (z_w - D_w)^2\right],$$

$$q = k\frac{D^2 - (z_w - D_w)^2}{2(L - l)}.$$

Equating the flow rates of two zones:

$$l = \frac{2LD(z_w - D)}{2D(z_w - D) + D^2 - (z_w - D_w)^2}.$$

Replacing $l$ in the flow rate equation:

$$q = \frac{k}{2L}\left[2D(z_w - D) + D^2 - (z_w - D_w)^2\right].$$

(b) Substituting numerical values for the case with $D_w = 2$ m:

$$q = \frac{4.0 \times 10^{-5} \times 3600 \times 24}{2 \times 100.0}\left[2 \times 5.0(7.0 - 5.0) + 5.0^2 - (7.0 - 2.0)^2\right] = 0.346 \text{ m}^3\text{/day}.$$

(c) In this case $D_w = 4$ m:

$$q = \frac{4.0 \times 10^{-5} \times 3600 \times 24}{2 \times 100.0}\left[2 \times 5.0(7.0 - 5.0) + 5.0^2 - (7.0 - 4.0)^2\right] = 0.622 \text{ m}^3\text{/day}.$$

Problem 3.5

A well of diameter 0.3 m is constructed to the full depth of an unconfined aquifer of thickness of 150 m. The water table is 10 m below the ground surface. A pumping test of 12 m$^3$/hour has resulted a drawdown of 10 m. Assuming $r_o$ = 400 m; calculate the coefficient of permeability of the aquifer. If the flow rate increases to 18 m$^3$/hour, and in the absence of any other data, what will be the best estimate for the drawdown in the well?

Solution:

In the unconfined aquifer the piezometric level $z$ (measured from the bottom of the well), in terms of $r$ (the distance of any point from the centre line of the well) and an observation well data (or a known point), may be expressed in the following form:

$$z = \sqrt{z_1^2 + \frac{q}{\pi k} \ln \frac{r}{r_1}}$$

(3.7)

For $r = 0.3 / 2 = 0.15$ m, $z = 150.0 - 10.0 - 10.0 = 130.0$ m.
The radius of influence $r_o$ defines the point(s) where the water level fully recovers to its original value; thus for $r = 400.0$ m, $z = 150.0 - 10.0 = 140.0$ m.
Substituting the above numerical values in Equation 3.7:

$$130.0 = \sqrt{140.0^2 + \frac{12.0}{3600 \times \pi k} \ln \frac{0.15}{400.0}} \rightarrow k = 3.1 \times 10^{-6} \text{ m/s}.$$

For $q = 18$ m$^3$/hour:

$$z = \sqrt{140.0^2 + \frac{18.0}{3600 \times \pi \times 3.1 \times 10^{-6}} \ln \frac{0.15}{400.0}} = 124.7 \text{ m}.$$

$$D_w = 150.0 - 10.0 - 124.7 = 15.3 \text{ m}.$$

Problem 3.6

A pumping test carried out in a 50 m thick confined aquifer resulted in a flow rate of 600 l/min. The thickness of the impermeable layer above the aquifer is 20 m and the original water level in the well was 2 m below the ground surface (which is also the top of the impermeable layer). Drawdowns in two observation wells located 50 m and 100 m from the well are 3 and 1 m respectively. Calculate:
(a) the coefficient of permeability of the aquifer,
(b) the drawdown in the well,
(c) the radius of influence. The diameter of the well is 0.6 m.

Solution:

In the confined aquifer the piezometric level $z$ (measured from the bottom of the well), in terms of $r$ (the distance of any point from the centre line of the well) and an observation well data (or a known point), is expressed in the following form:

$$z = z_1 + \frac{q}{2\pi Dk} \ln \frac{r}{r_1}$$

(3.8)

For $r = 50$ m, $z = 50.0 + 20.0 - 2.0 - 3.0 = 65.0$ m.
For $r = 100.0$ m, $z = 50.0 + 20.0 - 2.0 - 1.0 = 67.0$ m, therefore:

(a) $65.0 = 67.0 + \left[600.0/(1000 \times 60 \times 2\pi \times 50.0k)\right]\ln(50.0/100.0)$,

$k = 1.1 \times 10^{-5}$ m/s.

(b) $z = 67.0 + \left[600.0/(1000 \times 60 \times 2\pi \times 50.0 \times 1.1 \times 10^{-5})\right]\ln(0.3/100.0) = 50.2$ m,

$D_w = 50.0 + 20.0 - 2.0 - 50.2 = 17.8$ m.

(c) At $r = r_o$, $z = 50.0 + 20.0 - 2.0 = 68.0$ m; thus

$68.0 = 67.0 + \left[600.0/(1000 \times 60 \times 2\pi \times 50.0 \times 1.1 \times 10^{-5})\right]\ln(r_o/100.0)$,

$r_o = 141.3$ m.

## Problem 3.7

A soil profile consists of three layers with the properties shown in the table below. Calculate the equivalent coefficients of permeability parallel and normal to the stratum.

| Layer | Thickness $z$ (m) | $k_x$ (parallel, m/s) | $k_z$ (normal, m/s) |
|-------|-------------------|-----------------------|---------------------|
| 1 | 3.0 | $2.0 \times 10^{-6}$ | $1.0 \times 10^{-6}$ |
| 2 | 4.0 | $5.0 \times 10^{-8}$ | $2.5 \times 10^{-8}$ |
| 3 | 3.0 | $3.0 \times 10^{-5}$ | $1.5 \times 10^{-5}$ |

For the flow parallel to the layers:

$$k_x = \frac{z_1 k_{x1} + z_2 k_{x2} + \cdots + z_n k_{xn}}{z} \tag{3.9}$$

where $z = z_1 + z_2 + \cdots + z_n$.

$$k_x = \frac{3.0 \times 2.0 \times 10^{-6} + 4.0 \times 5.0 \times 10^{-8} + 3.0 \times 3.0 \times 10^{-5}}{3.0 + 4.0 + 3.0} = 9.6 \times 10^{-6} \text{ m/s.}$$

For the flow normal to the layers:

$$k_z = \frac{z}{z_1/k_{z1} + z_2/k_{z2} + \cdots + z_n/k_{zn}} \tag{3.10}$$

$$k_z = \frac{3.0 + 4.0 + 3.0}{3.0/1.0 \times 10^{-6} + 4.0/2.5 \times 10^{-8} + 3.0/1.5 \times 10^{-5}} = 6.1 \times 10^{-8} \text{ m/s.}$$

## Problem 3.8

For the sheet pile system shown in Figure 3.3(a), calculate the flow rate in $m^3$/day by constructing the flow net in the following two conditions:

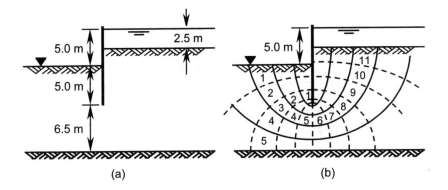

Figure 3.3. Problem 3.8: (a) sheet pile system, (b) the flow net under the sheet pile.

(a) $k_x = k_z = 5.0 \times 10^{-5}$ m/s, and

(b) $k_x$(horizontal) $= 5.0 \times 10^{-5}$ m/s, $k_z$(vertical) $= 3.0 \times 10^{-5}$ m/s.

Solution:

The flow net is constructed by sketching, as illustrated in Figure 3.3(b). The flow rate is expressed by:

$$q = kh\frac{N_f}{N_d} \qquad (3.11)$$

where $h$ is the total loss due to the seepage, $N_d$ is the number of equal drops in total head and $N_f$ is the total number of flow lanes.
From Figure 3.3(b):

$N_f = 5$, $N_d = 11$; therefore:

(a) $q = \dfrac{khN_f}{N_d} = \dfrac{(5.0 \times 10^{-5} \times 5.0 \times 5)3600 \times 24}{11}$,

$q = 9.82$ m$^3$/day/metre run.

(b) The material can be treated as an isotropic soil by assuming an equivalent isotropic coefficient of permeabilty of:

$$k_i = \sqrt{k_x k_z} \qquad (3.12)$$

and using a transformed scale of:

$$x' = x\sqrt{\frac{k_z}{k_x}} \qquad (3.13)$$

As the thickness of the sheet pile is ignored and there is no horizontal element (e.g. horizontal impermeable layer) thus there is no need to use the transformed scale and the flow net is the same as of case (a).
Using Equation 3.12:

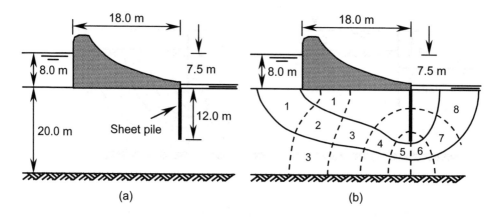

Figure 3.4. Problem 3.9: (a) concrete dam and sheet pile system, (b) the flow net.

$$k_i = \sqrt{5.0\times10^{-5}\times3.0\times10^{-5}} = 3.87\times10^{-5} \text{ m/s}.$$

$$q = \frac{k_i h N_f}{N_d} = \frac{(3.87.0\times10^{-5}\times5.0\times5)3600\times24}{11},$$

$q = 7.60$ m$^3$/day/metre run.

Problem 3.9

A concrete dam retains 8 m of water, as shown in Figure 3.4(a). Calculate the flow rate in m$^3$/day by constructing the flow net under the dam.

$k = 5 \times 10^{-5}$ m/s.

Solution:

The flow net is constructed by sketching and shown in Figure 3.4(b) from which:
$N_f = 3$, $N_d = 8$; therefore:

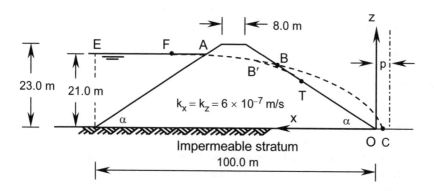

Figure 3.5. Problems 3.10 & 3.11.

$$q = \frac{khN_f}{N_d} = \frac{(5.0 \times 10^{-5} \times 7.5 \times 3)3600 \times 24}{8},$$

$q = 12.15 \text{ m}^3/\text{day/metre run}.$

Problem 3.10

For the earth dam section shown in Figure 3.5, calculate the flow rate in m$^3$/day.

Solution:

A traditional (Casagrande) method defines the phreatic surface as a parabola with its focus located at the origin $O$ of the $xz$ coordinate system (Figure 3.5). This parabola referred to as basic parabola can be defined mathematically if the coordinates of one point on the boundary (or within the seepage zone) is known. Experimental investigations have shown that the intersection point of the basic parabola and the water surface, point $F$, is located such that $FA = 0.3EA$, which means that the coordinates of point $F$ are known. The parabola has to be corrected at point $A$ to meet the requirements of the entry conditions. The distance of any point on the basic parabola from the focus is equal to the distance of this point from the directrix, which is located at an unknown distance of $p$ from the $z$-axis. This condition yields the equation of basic parabola as:

$$x = \frac{z^2 - p^2}{2p} \tag{3.14}$$

By substituting the coordinates of point $F$ in Equation 3.14, the value of $p$ so obtained is:

$$p = \sqrt{x_F^2 + z_F^2} - x_F \tag{3.15}$$

The flow rate is estimated by constructing the flow net schematically. An alternative solution is to assume a constant hydraulic gradient in the vertical sections:

$$q = Aki = (z \times 1)k\frac{dz}{dx}.$$

From Equation 3.14, $\dfrac{dz}{dx} = \dfrac{p}{z}$ and:

$$q = kp \tag{3.16}$$

In the absence of a toe drain the basic parabola intersects the downstream face at point $B$ as shown in Figure 3.5. In reality, the phreatic surface must be tangent to the downstream face at point $T$ with a distance $a$ from the origin $O$. In the Casagrande method, the correction length of $BT$ expressed by $\Delta a$ is found from experimental results. The distance $OB = \Delta a + a$ can be easily established, because the equations of the basic parabola and the downstream face are both known.

An alternative solution, called the Dupuit method gives the distance of $a$ as:

$$a = \frac{x_F}{\cos \alpha} - \sqrt{\frac{x_F^2}{\cos^2 \alpha} - \frac{z_F^2}{\sin^2 \alpha}} \tag{3.17}$$

In this case it is assumed that the correction curve passes through point $F$ and is tangent to the downstream face, thus the flow rate can be obtained by the following equation, which is different from Equation 3.16:

$$q = ka\sin\alpha\tan\alpha \tag{3.18}$$

Calculate the slope of the dam (equal for both upstream and downstream sides):

$$\tan\alpha = \frac{23.0}{(100.0-8.0)/2} = 0.5,$$

$\alpha = 26.56°.$

Referring to Figure 3.5:

$$EA = \frac{21.0}{\tan 26.56°} = 42.0 \text{ m},$$

$x_F = 100.0 - 42.0 + 0.3\times 42.0 = 70.6 \text{ m},$

$z_F = 21.0 \text{ m}.$

Using Equation 3.17:

$$a = \frac{70.6}{\cos 26.56°} - \sqrt{\frac{70.6^2}{\cos^2 26.56°} - \frac{21.0^2}{\sin^2 26.56°}},$$

$a = 15.49 \text{ m}.$

The flow rate is calculated from Equation 3.18:

$$q = 6.0\times 10^{-7}\times 15.49\times\sin 26.56°\times\tan 26.56°\times 3600\times 24 = 0.18 \text{ m}^3/\text{day/metre run}.$$

Problem 3.11

For the earth dam section shown in Figure 3.5, calculate the minimum length of the toe drain required to ensure that the phreatic surface becomes tangent to the downstream face.

Solution:

The equation of the downstream slope in the $zOx$ coordinate system is:

$$x = \frac{z}{\tan\alpha}.$$

When the point $O$ is located inside of the dam having a distance $L$ (the length of filter) from the previous origin; the equation of the downstream slope is modified to:

$$x = \frac{z}{\tan\alpha} - L.$$

To find the intersection point of this line and the basic parabola, we substitute the $x$ value from the equation of the line into equation of the basic parabola (Equation 3.14):

$$\frac{z^2 - p^2}{2p} = \frac{z}{\tan\alpha} - L.$$

Rearranging the above equation we obtain:

$$z^2 + \frac{-2p}{\tan \alpha} z + (2pL - p^2) = 0.$$

This equation is in the general form of:

$$az^2 + bz + c = 0,$$

which will yield two answers for $z$.

If the basic parabola is tangent to the downstream slope, these two answers are equal.

Thus

$$b^2 - 4ac = 0; \text{ or:}$$

$$(\frac{-2p}{\tan \alpha})^2 - 4(2pL - p^2) = 0,$$

$$L = \frac{p(1 + \cot^2 \alpha)}{2}.$$

The new $x$ coordinate of point $F$ (on the basic parabola) is:

$$x_F = 70.6 - L.$$

Substituting this value in Equation 3.15:

$$p = \sqrt{(70.6 - L)^2 + 21.0^2} - (70.6 - L),$$

The basic parabola is tangent to the downstream face if: $L = p(1 + \cot^2 \alpha)/2$, therefore:

$$L = \frac{p(1 + \cot^2 \alpha)}{2} = \frac{p(1 + 2.0^2)}{2} = 2.5p,$$

$p = 0.4L$. Thus

$$0.4L = \sqrt{(70.6 - L)^2 + 21.0^2} - (70.6 - L),$$

$$(70.6 - L)^2 + 441 = (70.6 - 0.6L)^2,$$

$$0.64L^2 - 56.48L + 441 = 0,$$

$$L = 8.65 \text{ m.}$$

## 3.3   REFERENCES AND RECOMMENDED READINGS

AS 2368.1990. *Test pumping of water wells*. NSW, Australia: Standard Association of Australia.

AS 1289.6.7.1. 1999. *Methods of testing soils for engineering purposes: Soil strength and consolidation tests-Determination of permeability of a soil-constant head method for a remoulded specimen*. NSW, Australia: Standard Association of Australia.

AS 1289.6.7.2. 1999. *Methods of testing soils for engineering purposes: Soil strength and consolidation tests-Determination of permeability of a soil-falling head method for a remoulded specimen*. NSW, Australia: Standard Association of Australia.

AS 1289.6.7.3. 1999. *Methods of testing soils for engineering purposes: Soil strength and consolidation tests-Determination of permeability of a soil-constant head method using a flexible wall permeameter.* NSW, Australia: Standard Association of Australia.

ASTM D-5084. 2000. *Standard test methods for measurement of hydraulic conductivity of saturated porous materials using a flexible wall permeameter.* West Conshohocken, PA: American Society for Testing and Materials.

ASTM D-2434-68. 2000. *Standard test method for permeability of granular soils (constant head).* West Conshohocken, PA: American Society for Testing and Materials.

ASTM D-4044. 1996. *Standard test method for (field procedure) for instantaneous change in head (Slug): Tests for determining hydraulic properties of aquifers.* West Conshohocken, PA: American Society for Testing and Materials.

Aysen, A. 2002. *Soil mechanics: Basic concepts and engineering applications.* Lisse: Balkema.

BS 5930. 1981. *Code of practice for site investigation.* London: British Standard Institution.

Head, K.H. 1986. *Manual of soil laboratory testing.* London: Pentech Press.

Roberson, J.A., Cassidy, J.J. & Chaudhry, M.H. 1997. *Hydraulic engineering.* New York: John Wiley & Sons.

# CHAPTER 4

# Shear Strength of Soils and Failure Criteria

## 4.1 INTRODUCTION

The problems solved in this chapter describe the *shear strength characteristics* of soils and related *failure criteria*. Problems 4.1 to 4.3 show the applications of the stress transformation equations in soils. The *shear strength parameters*, which include cohesion and internal friction angle, are investigated through Problems 4.4 to 4.9. Both, the effective stress analysis and the total stress analysis, are applied. The post peak behaviour of the soil may be predicted by using the *critical state models* that are now common for specific types of soils. The practical applications of the critical state theory are described by Problems 4.10 to 4.12.

## 4.2 PROBLEMS

Problem 4.1

At a point 15 m below the ground surface, the relationship between the effective vertical stress $\sigma'_z$ and the effective lateral stress $\sigma'_x$ is: $\sigma'_x = \sigma'_z (1 - \sin\phi')$. If the water table is 2 m below the ground surface, calculate the normal and shear stresses on the two perpendicular planes $P$ and $Q$ (Figure 4.1(a)) where the angle $\alpha$ for the $P$ plane is $45° + \phi' / 2$. $c' = 0$, $\phi' = 40°$, $\rho_{dry} = 1.7$ Mg/m$^3$ and $\rho_{sat} = 1.95$ Mg/m$^3$.

Solution:

At $z = 15$ m:

$\sigma_z = 1.7 \times 9.81 \times 2.0 + 1.95 \times 9.81 \times 13.0 = 282.0$ kPa,

$u = 1.0 \times 9.81 \times 13.0 = 127.5$ kPa, $\sigma'_z = 282.0 - 127.5 = 154.5$ kPa.

$\sigma'_x = \sigma'_z (1 - \sin\phi') = 154.5(1 - \sin 40.0°) = 55.2$ kPa. Due to symmetry $\tau_{xz} = 0$.

The normal and shear stresses on any plane $P$, with angle $\alpha$ to the $x$-axis, are determined by the following equations:

$$\sigma' = \frac{\sigma'_z + \sigma'_x}{2} + \frac{\sigma'_z - \sigma'_x}{2}\cos 2\alpha + \tau_{xz}\sin 2\alpha \qquad (4.1)$$

$$\tau = \frac{\sigma'_z - \sigma'_x}{2}\sin 2\alpha - \tau_{xz}\cos 2\alpha \qquad (4.2)$$

For the plane $P$, $\alpha = 45.0° + 40.0°/2 = 65.0° \rightarrow 2\alpha = 130.0°$. Thus

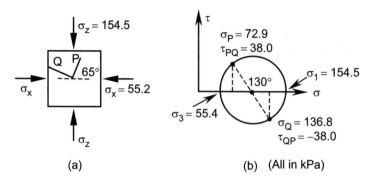

Figure 4.1. Problem 4.1.

$\sigma = (154.5 + 55.2)/2 + [(154.5 - 55.2)/2]\cos 130.0° = 72.9$ kPa.

$\tau = [(154.5 - 55.2)/2]\sin 130.0° = 38.0$ kPa.

For the plane $Q$, $\alpha = 90.0° + 45.0° + 40.0°/2 = 155.0° \to 2\alpha = 310°$ or $2\alpha = -50.0°$.

$\sigma = (154.5 + 55.2)/2 + [(154.5 - 55.2)/2]\cos(-50.0°) = 136.8$ kPa.

$\tau = [(154.5 - 55.2)/2]\sin(-50.0°) = -38.0$ kPa.

The results are illustrated in Figure 4.1(b).

Problem 4.2

At a point within a soil mass, the effective lateral and shear stresses are 100 kPa and 50 kPa respectively. Calculate the effective vertical stress to cause the failure of the point. $c' = 0$, $\phi' = 30°$.

Solution:

The Mohr-Coulomb failure criterion can be expressed in terms of Cartesian stresses by enforcing the Mohr' circle of stress (Equations 4.1 and 4.2) to be tangent to the failure envelope (the concept of failure in the form of $\tau = \sigma' \tan\phi' + c'$):

$$(\sigma'_z - \sigma'_x)^2 + (2\tau_{xz})^2 = [2c'\cos\phi' + (\sigma'_z + \sigma'_x)\sin\phi']^2 \qquad (4.3)$$

Which means, the radius of the Mohr's circle of stress, (square root of the left term) is equal to the distance of the centre of the circle from the failure envelope (square root of the right term).

Substituting the numerical values in the above equation and noting that:
$\sigma_z > \sigma_x = 100$ kPa:

$(\sigma'_z - 100.0)^2 + (2 \times 50.0)^2 = [(\sigma'_z + 100.0)\sin 30.0°]^2$. After simplifying and rearranging:

$0.75(\sigma'_z)^2 - 250.00\sigma'_z + 17500.0 = 0$,

$\sigma'_z = 233.3$ kPa.

Problem 4.3

An element of soil in $x$-$z$ plane is subjected to $\sigma'_z$, $\sigma'_x$, and $\tau_{xz}$. Assuming that: $\tau_{xz} = 0.306\,\sigma'_x$, calculate the ratio $\sigma'_x / \sigma'_z$ at failure. $c' = 0$, $\phi' = 36°$.

Solution:

Substituting the given data in the failure criterion expressed by Equation 4.3:

$$(\sigma'_z - \sigma'_x)^2 + (2 \times 0.306\sigma'_x)^2 = [(\sigma'_z + \sigma'_x)\sin 36.0°]^2.$$

After rearrangements:

$$1.029(\sigma'_x / \sigma'_z)^2 - 2.691(\sigma'_x / \sigma'_z) + 0.6545 = 0.$$

$$\sigma'_x / \sigma'_z = 0.271, 2.344.$$

Both answers are acceptable, however practically $\sigma'_x / \sigma'_z$ is smaller than 1.

Problem 4.4

The results of a direct shear test on a specimen of dry sand are as follows:
Normal stress = 96.6 kPa; shear stress at failure = 67.7 kPa. By means of a Mohr's circle of stresses, find the magnitude and directions of the principal stresses acting on a soil element within the zone of failure.

Solution:

The equation of the failure envelope is:

$\tau = \sigma' \tan\phi' + c'$, therefore: $67.7 = 96.6 \tan\phi' + c'$.

For dry sand assume $c' = 0$; thus

$$\tan\phi' = \frac{67.7}{96.6} = 0.700 \rightarrow \phi' = 35.0°.$$

The normal and shear stresses on any plane $P$ may be expressed in terms of principal stresses and the angle of the plane ($\alpha$) from the major principal stress plane. Thus rearranging Equations 4.1 and 4.2:

$$\sigma' = \frac{\sigma'_1 + \sigma'_3}{2} + \frac{\sigma'_1 - \sigma'_3}{2}\cos 2\alpha \qquad (4.4)$$

$$\tau = \frac{\sigma'_1 - \sigma'_3}{2}\sin 2\alpha \qquad (4.5)$$

If $P$ becomes the failure plane, then, $2\alpha = \pi / 2 + \phi'$:

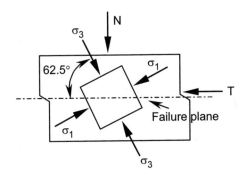

Figure 4.2. Problem 4.4.

$$\sigma' = \frac{\sigma_1' + \sigma_3'}{2} - \frac{\sigma_1' - \sigma_3'}{2} \sin \phi' \qquad (4.6)$$

$$\tau = \frac{\sigma_1' - \sigma_3'}{2} \cos \phi' \qquad (4.7)$$

Substituting the given data in the above equations:

$$96.6 = \frac{\sigma_1' + \sigma_3'}{2} - \frac{\sigma_1' - \sigma_3'}{2} \sin 35.0°, \quad 67.7 = \frac{\sigma_1' - \sigma_3'}{2} \cos 35.0°.$$

Solving for $\sigma_1'$ and $\sigma_3'$ we obtain:

$\sigma_1' = 226.7$ kPa, $\sigma_3' = 61.4$ kPa.

$\alpha = (90.0° + 35.0°)/2 = 62.5°$.

The results are shown in Figure 4.2.

Problem 4.5

Data obtained from a drained triaxial test are as follows:

| Test no. | $\sigma_3$ (kPa) | $\sigma_1 - \sigma_3$ at peak (kPa) |
|---|---|---|
| 1 | 50 | 191 |
| 2 | 100 | 226 |
| 3 | 150 | 261 |

Determine the drained shear strength parameters.

Solution:

Test 1:

In drained test $\sigma_3' = \sigma_3 = 50.0$ kPa,

$\sigma_1' = 50.0 + 191.0 = 241.0$ kPa.

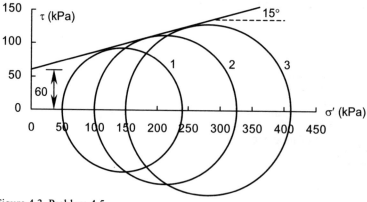

Figure 4.3. Problem 4.5.

Test 2:

$\sigma_3' = 100.0$ kPa,

$\sigma_1' = 100.0 + 226.0 = 326.0$ kPa.

Test 3:

$\sigma_3' = 150.0$ kPa,

$\sigma_1' = 150.0 + 261.0 = 411.0$ kPa.

Mohr's circles of stress are shown in Figure 4.3.
The best estimation for drained shear strength parameters are:
$c' = 60$ kPa and $\phi' = 15°$.

Problem 4.6

The results of three consolidated-undrained triaxial tests on identical specimens of a particular soil are:

| Test no. | $\sigma_3$ (kPa) | $\sigma_1 - \sigma_3$ at peak (kPa) | $u$ at peak (kPa) |
|----------|---------|-----------------------|-----------------|
| 1 | 200 | 244 | 55 |
| 2 | 300 | 314 | 107 |
| 3 | 400 | 384 | 159 |

Determine $c'$ and $\phi'$.
What would be the expected pore pressure at failure for a test with $\sigma_3 = 100$ kPa?

Solution:

Calculate $\sigma'_3$ and $\sigma'_1$ at failure for each test:

Test 1:

$\sigma_3' = 200.0 - 55.0 = 145.0$ kPa, $\sigma_1' = 145.0 + 244.0 = 389.0$ kPa.

Test 2:

$\sigma_3' = 300.0 - 107.0 = 193.0$ kPa, $\sigma_1' = 193.0 + 314.0 = 507.0$ kPa.

Test 3:

$\sigma_3' = 400.0 - 159.0 = 241.0$ kPa, $\sigma_1' = 241.0 + 384.0 = 625.0$ kPa.

Mohr's circles of stress are shown in Figure 4.4 from which:
$c' = 10$ kPa and $\phi' = 25°$.
To calculate the shear strength parameters related to the total stresses:

Test 1:

$\sigma_3 = 200.0$ kPa, $\sigma_1 = 200.0 + 244.0 = 444.0$ kPa.

Test 2:

$\sigma_3 = 300.0$ kPa, $\sigma_1 = 300.0 + 314.0 = 614.0$ kPa.

Test 3:

$\sigma_3 = 400.0$ kPa, $\sigma_1 = 400.0 + 384.0 = 784.0$ kPa.

Mohr's circles of stress are shown in Figure 4.4 (dashed circles) from which: $c = 40$ kPa and $\phi = 15°$.

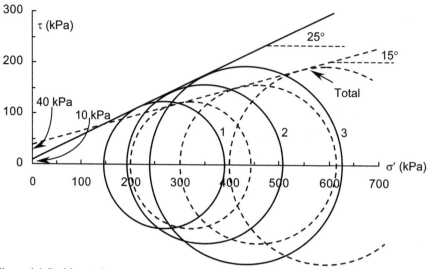

Figure 4.4. Problem 4.6.

To obtain the failure criterion in terms of principal stresses, the normal and shear stresses on the plane of failure expressed by Equations 4.6 and 4.7 are substituted in the equation of failure envelope: $\tau = \sigma' \tan\phi' + c'$.

Alternatively Cartesian stresses in Equation 4.3 are replaced by principal stresses and the Cartesian shear stress is set to zero:

$$\sigma_1' = \sigma_3' \tan^2(45° + \phi'/2) + 2c' \tan(45° + \phi'/2) \qquad (4.8)$$

The above equation may also be presented in the following form:

$$\sigma_3' = \sigma_1' \tan^2(45° - \phi'/2) - 2c' \tan(45° - \phi'/2) \qquad (4.9)$$

For $\sigma_3 = 100$ kPa use Equation 4.8 for both total and effective stresses.

For total stresses:

$$\sigma_1 = 100.0 \tan^2(45.0° + 15.0°/2) + 2 \times 40.0 \tan(45.0° + 15.0°/2) = 274.1 \text{ kPa}.$$

For effective stresses:

$$(274.1 - u) = (100.0 - u) \tan^2(45.0° + 25.0°/2) + 2 \times 10.0 \tan(45.0° + 25.0°/2),$$

$$u = 2.6 \text{ kPa}.$$

Problem 4.7

The results of drained and consolidated-undrained triaxial tests on two samples of normally consolidated clay are shown below:

| Type of the test | $\sigma_3$ (kPa) | $\sigma_1 - \sigma_3$ at peak (kPa) |
|---|---|---|
| Drained | 300 | 650 |
| Consolidated-undrained | 200 | 250 |

Determine:
(a) $\phi'$ from the drained test,
(b) $\phi$ from the consolidated-undrained test,
(c) the pore pressure in the consolidated-undrained test at failure.

Solution:

(a) For normally consolidated clay $c' = 0$, thus using Equation 4.8:

$$\sigma_1' = \sigma_3' \tan^2(45° + \phi'/2), (300.0 + 650.0) = 300.0 \tan^2(45.0° + \phi'/2),$$

$$\tan^2(45.0° + \phi'/2) = 3.1667 \rightarrow \phi' = 31.33° \approx 31.3°.$$

(b) Using the same equation for total stresses and assuming $c = 0$:

$$(200.0 + 250.0) = 200.0 \tan^2(45.0° + \phi/2),$$

$$\tan^2(45.0° + \phi/2) = 2.25 \rightarrow \phi = 22.62° \approx 22.6°.$$

(c) Substitute the results of the consolidated-undrained test in the equation of the failure criterion using $\phi'$ value obtained in part (a):

$$(200.0 + 250.0 - u) = (200.0 - u)\tan^2(45.0° + 31.33°/2) = (200 - u)3.166; \text{ thus}$$

$$u = 84.6 \text{ kPa}.$$

The results are graphically illustrated in Figure 4.5.


Problem 4.8

A soil has the following properties: $n$ (porosity) $= 0.38$, $E_s$ (Modulus of Elasticity) $= 10$MPa, $\mu = 0.3$. The bulk modulus of the pore water is 2200 MPa. Estimate the pore pressure coefficient $B$.

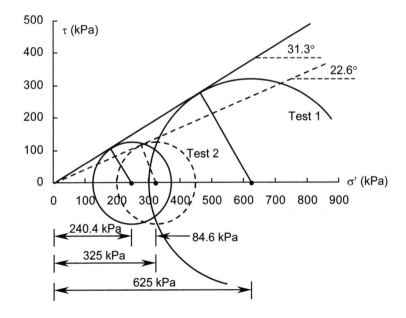

Figure 4.5. Problem 4.7.

Solution:

If an element of saturated soil is subjected to isotropic compression $\Delta\sigma$, the instant increase in pore pressure $\Delta u$ (excess pore pressure) is calculated from:

$$\Delta u = \frac{1}{1 + \frac{nE_s}{3K(1-2\mu)}}\Delta\sigma = B\Delta\sigma \qquad (4.10)$$

where $n$ is the porosity, $E_s$ is the Modulus of Elasticity of soil, $\mu$ is the Poisson's ratio, K is the bulk modulus of the pore water and $B$ is termed the pore pressure coefficient. Substituting the given data in the above equation:

$$B = \frac{1}{1 + \frac{nE_s}{3K(1-2\mu)}} = \frac{1}{1 + \frac{0.38 \times 10.0}{3 \times 2200.0(1-2\times 0.3)}} = 0.9986.$$

Problem 4.9

An unconfined compression test has given a *UCS* value of 126.6 kPa. The effective shear strength parameters are: $c' = 25$ kPa, $\phi' = 30°$. Assuming the pore pressure parameter $A = -0.09$, calculate the initial pore pressure in the sample.

Solution:

Using the failure criterion in terms of effective stresses (Equation 4.8):

$$\sigma_1' = \sigma_3' \tan^2(45.0° + 30.0°/2) + 2 \times 25.0 \tan(45.0° + 30.0°/2),$$

$$\sigma_1' = 3\sigma_3' + 86.6.$$

The unconfined compression strength represents:

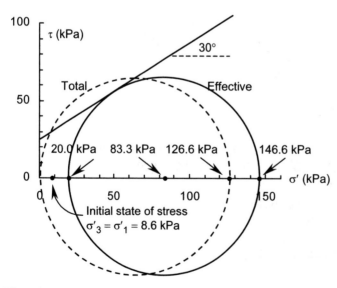

Figure 4.6. Problem 4.9.

$\sigma_1 - \sigma_3 = \sigma_1 - 0.0 = \sigma_1' - \sigma_3'$. Thus

$\sigma_1' - \sigma_3' = 126.6 \rightarrow \sigma_1' = \sigma_3' + 126.6$.

Substituting $\sigma'_1$ in the equation of the failure criterion:

$\sigma_3' + 126.6 = 3\sigma_3' + 86.6 \rightarrow \sigma_3' = 20.0$ kPa.

The effective major principal stress is:

$\sigma_1' = \sigma_3' + 126.6 = 20.0 + 126.6 = 146.6$ kPa.

In unconfined compression test the confining pressure $\sigma_3$ is zero:

$\sigma_3 = \sigma_3' + u = 20.0 + u = 0 \rightarrow u = -20.0$ kPa.

The excess pore pressure in a traditional triaxial compression test is determined from the following equation where $A$ and $B$ are termed pore pressure coefficients:

$$\Delta u = B\left[\Delta\sigma_3 + A(\Delta\sigma_1 - \Delta\sigma_3)\right] \tag{4.11}$$

In an unconfined compression test $\Delta\sigma_3 = 0$. Furthermore the incremental stresses can be replaced by $\sigma_1$ and $\sigma_3$ for evaluation of the average values for $A$ and $B$. Assuming $B = 1$ and knowing that:

$\sigma_1 - \sigma_3 = \sigma_1' - \sigma_3'$,

the pore pressure coefficient $A$ at failure is calculated from:

$$A = \frac{\Delta u}{\sigma_1 - \sigma_3} = \frac{\Delta u}{\sigma_1' - \sigma_3'}, \text{ thus}$$

$$A = -0.09 = \frac{\Delta u}{\sigma_1 - \sigma_3} = \frac{\Delta u}{126.6} \rightarrow \Delta u = -11.4 \text{ kPa}.$$

The initial pore pressure is:

$u_i = -20.0 - (-11.4) = -8.6$ kPa.

The results are graphically presented in Figure 4.6.


Problem 4.10

The values of the critical state parameters for a particular type of clay are:
$N = 2.1$, $\lambda = 0.087$, $\Gamma = 2.05$, $M = 0.95$.

Two samples of this soil are consolidated under a confining pressure of 300 kPa. One sample has been subjected to a drained triaxial test whilst the second sample has been sheared in an undrained condition. Determine:

(a) the deviator stress at the critical state for both the drained and undrained tests,
(b) the pore pressure in the undrained test at the critical state,
(c) the volumetric strain in the drained test when the sample approaches the critical state.

Solution:

(a) Projection of the critical state line (*CSL*) onto the $p'$-$q'$ plane defines the state of stress at the critical state and is a line that passes through the origin with gradient $M$.

$$q' = Mp' \tag{4.12}$$

where $q' = \sigma_1' - \sigma_3'$ and $p' = \dfrac{\sigma_1' + 2\sigma_3'}{3}$.

The initial state of the sample is point $C$ in Figure 4.7(a). It can be shown that the stress path of the drained test on the $p'$-$q'$ plane has a slope of 3 vertical to 1 horizontal (line $CD$). The stress path for the undrained test is represented by $CU$. Points $D$ and $U$ that are located on the critical state line represent the states of the two samples at critical state. For the drained test the equation of the drained path (line $CD$ in Figure 4.7(a)) is:

$$q' = 3(p' - 300.0).$$

This ensures the slope of 3 vertical, 1 horizontal on the $p'$-$q'$ plane. Substituting $q'$ in the equation of *CSL*:

$$q' = 0.95 p' = 3(p' - 300.0), \text{ thus}$$

$p'$(at critical state) = 439.0 kPa.

Substituting the above value of $p'$ in Equation 4.12:

$$q' = \sigma'_1 - \sigma'_3 = 0.95 \times 439.0 = 417.0 \text{ kPa}.$$

The state of the sample subjected to an isotropic compression or confining pressure $\sigma_3$ is defined by a relationship between $v$ and $p'$ called the normal compression line (*NCL*), as shown in Figures 4.7(b) and 4.7(c). Specific volume $v$ is a dimensionless parameter and represents a volume in which the solids occupy a unit volume. From a phase diagram it can be shown that $v = 1 + e$ where $e$ is the void ratio. The normal compression line is established by means of a triaxial compression test. The equation of the normal compression line (*NCL*) in $v$, $\ln p'$ coordinate system is:

$$v = N - \lambda \ln p' \tag{4.13}$$

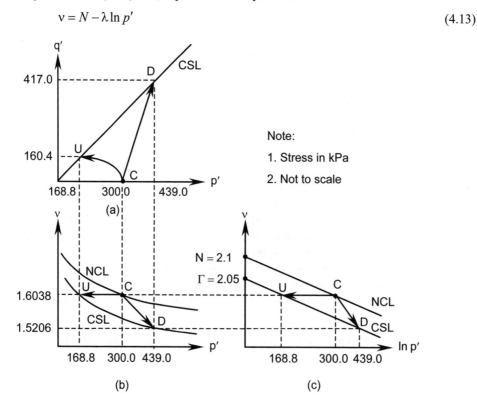

Figure 4.7. Problem 4.10.

The projection of the *CSL* on to the v, ln $p'$ coordinate system is approximated by a line parallel to the *NCL* (Figure 4.7(c)):

$$v = \Gamma - \lambda \ln p' \qquad (4.14)$$

For the undrained test we first substitute numerical values in the *NCL* equation and calculate the corresponding specific volume during the test and at failure:

$v = 2.1 - 0.087 \times \ln 300.0 = 1.6038.$

The magnitude of $p'$ at critical state is calculated by substituting the specific volume in the *CSL* equation:

$1.6038 = 2.05 - 0.087 \times \ln p' \rightarrow p' = 168.8 \, \text{kPa}.$

The deviator stress at the critical state is:

$q' = \sigma_1' - \sigma_3' = Mp' = 0.95 \times 168.8 = 160.4 \, \text{kPa}.$

(b) From the definition of $p'$ we have:

$p' = (\sigma_1' + 2\sigma_3')/3 = (\sigma_1' - \sigma_3' + \sigma_3' + 2\sigma_3')/3 = (q' + 3\sigma_3')/3,$

$p' = (160.4 + 3\sigma_3')/3 = 168.8,$

$\sigma_3' = 115.3 \, \text{kPa}.$

$u = 300.0 - 115.3 = 184.7 \, \text{kPa}.$

(c) Calculate the specific volume at critical state (drained test):

$v = \Gamma - \lambda \ln p' = 2.05 - 0.087 \ln 439.0 = 1.5206.$

The general definition of volumetric strain is expressed by:

$$\varepsilon_V = \frac{\Delta V}{V} \approx \varepsilon_1 + \varepsilon_2 + \varepsilon_3 \qquad (4.15)$$

where $\Delta V$ is the volume change, $V$ is the initial volume; $\varepsilon_1$, $\varepsilon_2$, and $\varepsilon_3$ are the axial strains in the direction of major principal stresses. The above equation can be expressed in terms of specific volume:

$$\varepsilon_V = \frac{\Delta v}{v} \qquad (4.16)$$

Thus $\varepsilon_V = (1.6038 - 1.5206)/1.6038 = 0.0519 = 5.19\%.$

The results are illustrated in Figure 4.7.

Problem 4.11

In a drained triaxial test carried out on a sample of the clay of Problem 4.10, the sample was first consolidated under a confining pressure of 400 kPa. It was then unloaded to 300 kPa and, after equilibrium was reached, it was sheared in drained conditions. If the $\kappa$ value is 0.037, calculate the volumetric strain at failure.

Solution:

The equation of expansion line (unloading) in v, ln $p'$ coordinate system (Figure 4.8(c)) is in the form:

$$v = v_\kappa - \kappa \ln p' \qquad (4.17)$$

where

$$\kappa = -\frac{v_C - v_{C_1}}{\ln p'_C - \ln p'_{C_1}} \qquad (4.18)$$

is the slope of the expansion line and $v_\kappa$ is the magnitude of $v$ at $p' = 1$ kPa. The parameter $v_\kappa$ is not a constant for the soil and its magnitude depends on the magnitude of $p'_{C_1}$.

Note in this example $p'_{C_1} = 400$ kPa, and $p'_C = 300$ kPa. Alternatively:

$$(v - v_C)/(\ln p' - \ln p'_C) = -\kappa, \text{ or: } (v - v_{C_1})/(\ln p' - \ln p'_{C_1}) = -\kappa \qquad (4.19)$$

Calculate the specific volume for $p' = 400.0$ kPa on the *NCL*:
$v_{400} = 2.1 - 0.087 \times \ln 400.0 = 1.5787$.

Calculate the specific volume after unloading to $p' = 300.0$ kPa:

$$\kappa = \frac{-(v_{300} - v_{400})}{(\ln 300.0 - \ln 400.0)} = 0.037, \quad \frac{-(v_{300} - 1.5787)}{(\ln 300.0 - \ln 400.0)} = 0.037 \rightarrow v_{300} = 1.5893.$$

From part (c) of Problem 4.10 the specific volume at critical state is 1.5206.
Volumetric strain from equilibrium conditions at $p' = 300.0$ kPa and $v = 1.5893$ to critical state at $p' = 439.0$ kPa and $v = 1.5206$ is:
$\varepsilon_V = (1.5893 - 1.5206)/1.5893$,
$\varepsilon_V = 0.0432 = 4.32\%$ (compression).

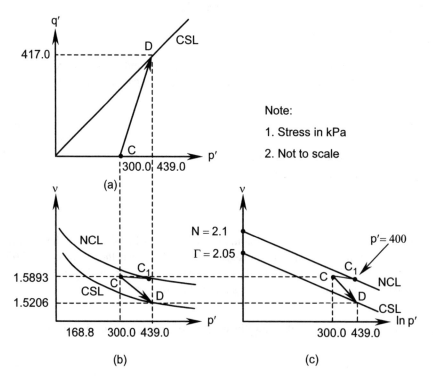

Figure 4.8. Problem 4.11.

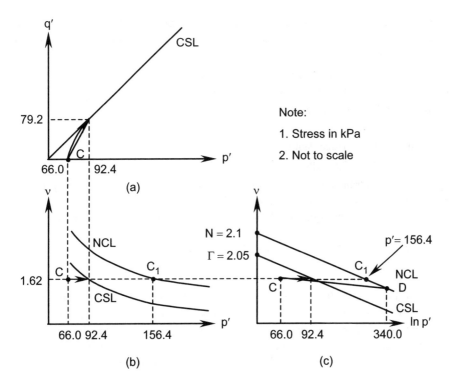

Figure 4.9. Problem 4.12.

This is less than 5.19% obtained in Problem 4.10 due to its lightly overconsolidated state. However the deviator stress at critical state is the same as the drained sample of Problem 4.10. Volumetric strain during unloading from $p' = 400.0$ kPa to $p' = 300.0$ kPa is:

$$\varepsilon_V = \frac{1.5787 - 1.5893}{1.5787},$$

$\varepsilon_V = -0.0067 = -0.67\%$ (expansion).

The behaviour of sample during the test is shown in Figure 4.8.

Problem 4.12

The critical state parameters of a soil are:

$M = 0.857$, $\lambda = 0.095$, $N = 2.1$, $\Gamma = 2.05$, $\kappa = 0.045$.

Specimens of this soil have been consolidated and unloaded to obtain an initial void ratio of 0.62.

(a) If the specimens are subjected to an undrained triaxial test, find the minimum overconsolidation ratio ($OCR = m$) above which the pore pressure at the critical state becomes negative,

(b) calculate the volumetric strains for three specimens of $OCR = 1$, $OCR = m$ (as defined above) and $OCR = 8$ that are subjected to drained triaxial tests.

Solution:

(a) We consider the case where the pore pressure at failure becomes zero. At this state the effective stress path and the total stress path intersect each other on the *CSL* ($p'$-$q'$ plane) so that the $q'$ is the same for both stress paths at this point (critical state).

At the critical state the specific volume is: $v = 1 + e = 1 + 0.62 = 1.62$.

Using Equation 4.14 the value of $p'$ at $v = 1.62$ becomes:

$$v = \Gamma - \lambda \ln p' = 1.62 = 2.05 - 0.095 \ln p',$$

$p' = 92.4$ kPa.

The deviator stress at the critical state is calculated from Equation 4.12:

$$q' = Mp' = 0.857 \times 92.4 = 79.2 \text{ kPa.}$$

The progress of the undrained triaxial test in terms of the total stresses, on the $p'$-$q'$ plane, is a line with the slope of 3 vertical to 1 horizontal.

Referring to Figure 4.9(a) the equation of the total stress path may be written as follows:

$$\frac{q - 79.2}{p - 92.4} = 3, \text{ which has a slope of 3 vertical, 1 horizontal.}$$

For $q = 0, p = p'_C = 66.0$ kPa.

This means an unloading (after consolidation) has yielded $p'_C = 66.0$ kPa as the initial state for triaxial test. In order to find the corresponding preconsolidation pressure (the isotropic compression where the sample was unloaded; point $D$ in Figure 4.9(c)) on the normal compression line we intersect this line with the expansion (or unloading) line. For this purpose first we define the slope of the expansion line (Equation 4.18):

$$\kappa = -\frac{v_C - v_D}{\ln p'_C - \ln p'_D} \rightarrow 0.045 = -\frac{1.62 - v_D}{\ln 66.0 - \ln p'_D}.$$

Point $D$ is located on the normal compression line; thus

$$v_D = N - \lambda \ln p' = 2.1 - 0.095 \ln p'_D.$$

Substituting $v_D$ in the equation of the expansion line:

$$0.045 = -\frac{1.62 - (2.1 - 0.095 \ln p'_D)}{\ln 66.0 - \ln p'_D}.$$

Solving for $p'_D$:

$p'_D = 340.0$ kPa.

The value of $v$ at this point is:

$$v_D = N - \lambda \ln p' = 2.1 - 0.095 \ln 340.0 = 1.5462.$$

By definition the overconsolidation ratio is the ratio of the preconsolidation pressure to the existing effective vertical stress on the element:

$OCR = 340.0 / 66.0 = 5.15$.

Having the values of specific volumes at points $D$ and $C$ we may use the following equation for determination of the overconsolidation ratio which is applicable for any linear unloading in $v$, $\ln p'$ coordinate system:

$$OCR = \exp\left(\frac{v_C - v_D}{\kappa}\right) \tag{4.20}$$

$$OCR = \exp\left(\frac{1.62 - 1.5462}{0.045}\right) = 5.15.$$

For the *OCR* values lower than 5.15 ($p'_D < 340.0$ kPa) the pore pressure at the critical state is positive; for the *OCR* values higher than 5.15 ($p'_D > 340.0$ kPa) the pore pressure is negative. The pore pressure can be calculated in a way similar to part (b) of Problem 4.10. As an example if $p'_D = 400.0$ kPa:

$$v_D = N - \lambda \ln p' = 2.1 - 0.095 \ln 400.0 = 1.5308.$$

Using the definition of the slope of the expansion line:

$$\kappa = -\frac{v_C - v_D}{\ln p'_C - \ln p'_D} \rightarrow 0.045 = -\frac{1.62 - 1.5308}{\ln p'_C - \ln 400},$$

$p'_C = 55.1$ kN $< 66.0$ kPa.

From the definition of $p'$ we have (part (b) of Problem 4.10):

$$p' = (\sigma'_1 + 2\sigma'_3)/3 = (\sigma'_1 - \sigma'_3 + \sigma'_3 + 2\sigma'_3)/3 = (q' + 3\sigma'_3)/3 = 92.4 \text{ kPa}.$$

$$p' = (\sigma'_1 + 2\sigma'_3)/3 = (79.2 + 3\sigma'_3)/3 = 92.4 \rightarrow \sigma'_3 = 66.0 \text{ kPa}.$$

$u = 55.1 - 66.0 = -10.9$ kPa.

The results of part (a) are shown in Figure 4.9.

(b) For *OCR* = 1:

$$v_D = 1.62 = N - \lambda \ln p' = 2.1 - 0.095 \ln p'_D,$$

$p'_D = 156.4$ kPa.

The $p'$ value at the critical state is:

$$q' = 3(p' - 156.4) = 0.857 p' \rightarrow p' = 218.9 \text{ kPa}.$$

The corresponding specific volume at the critical state:

$$v = \Gamma - \lambda \ln p' = 2.05 - 0.095 \ln 218.9 = 1.5381.$$

$$\varepsilon_V = \frac{(1.62 - 1.5381)}{1.62} = 0.0506 = 5.06\% \approx 5.1\% \text{ (compression)}.$$

For *OCR* = 5.15, $p' = 92.4$ kPa.

$$v = \Gamma - \lambda \ln p' = 2.05 - 0.095 \ln 92.4 = 1.62.$$

$$\varepsilon_V = \frac{(1.62 - 1.62)}{1.62} = 0.0 = 0.0\%.$$

For *OCR* = 8:

$$\kappa = -\frac{v_C - v_D}{\ln p'_C - \ln p'_D} \rightarrow 0.045 = -\frac{1.62 - v_D}{\ln(1/8)}.$$

$v_D = 1.5264.$

$$v_D = N - \lambda \ln p'_D = 1.5264 = 2.1 - 0.095 \ln p'_D \rightarrow p'_D = 419.0 \text{ kPa}.$$

$p'_C = 419.0/8 = 52.4$ kPa.

$$q' = 3(p' - 52.4) = 0.857 p',$$

$p' = 73.3$ kPa.

The corresponding specific volume at critical state:

$v = \Gamma - \lambda \ln p' = 2.05 - 0.095 \ln 73.3 = 1.6420$.

$\varepsilon_V = \dfrac{(1.62 - 1.642)}{1.62} = -0.0136 = -1.36\%$ (expansion).

## 4.3   REFERENCES AND RECOMMENDED READINGS

Arthur, J. R. F., & Aysen, A. 1977. Ruptured sand sheared in plane strain. *Proc. intern. conf. SMFE*, 1: 19-22. Tokyo.

Atkinson, J. H., & Bransby, P. L. 1978. *The mechanics of soils*. London: McGraw-Hill.

Aysen, A. 2002. *Soil mechanics: Basic concepts and engineering applications*. Lisse: Balkema.

Britto, A. & Gunn, M.J. 1987. *Critical state soil mechanics via finite elements*. Chichester, UK: Ellis Horwood.

Jewell, R. A. 1989. Direct shear tests on sands. *Geotechnique*, 39(2): 309-322.

Powrie, W. 1997. *Soil mechanics-concepts and applications*. London: E & FN Spon.

Roscoe, K. H.1953. An apparatus for the application of simple shear to soil samples. *Proc. 3rd intern. conf. SMFE*, 1: 186-191. Switzerland.

Roscoe, K.H. & Burland, J.B. 1968. On the generalized stress-strain behaviour of wet clay. In J. Heyman & F.A. Leckie (eds), *Engineering plasticity*: 535-639. Cambridge, UK: Cambridge University Press.

Terzaghi, K., Peck, R. B., & Mesri, G. 1996. *Soil mechanics in engineering practice*. 3rd edition. New York: John Wiley & Sons.

# CHAPTER 5

# Stress Distribution and Settlement in Soils

## 5.1  INTRODUCTION

The problems solved in this chapter are divided into three major categories:
1. *Stress distribution* within an idealized elastic soil mass due to applied external and internal loading (Problems 5.1 to 5.11).
2. Calculation of *elastic settlement* (Problems 5.12 to 5.16).
3. Calculation of *contact pressure* under the base and settlements through two different concepts of rigid and elastic footing (Problems 5.17 to 5.21).
In categories 1 and 2 the soil is assumed to be an ideal semi-infinite homogenous elastic material obeying Hooke's law, and a linear elastic stress-strain model is employed to determine the stress distribution within the soil. The elastic properties include the *Modulus of Elasticity $E_s$* and *Poisson's ratio* $\mu$. For soils where the compressibility characteristics are non-uniform and depend upon the state of the stress, the concept of a compressibility index $m_v$ is introduced. In this case, an average $m_v$ within a zone of influence is estimated by in-situ testing.

## 5.2  PROBLEMS

Problem 5.1

A cylinder of diameter 150 mm and height 300 mm is filled with sand. The surface of the sand is subjected to a vertical stress of 300 kPa causing 4 mm settlement under the loading plate. Calculate the lateral stress on the wall of the cylinder and Modulus of Elasticity of the sand. Poisson's ratio for the sand is 0.2.

Solution:

The average strain along the length of the sample is: $\varepsilon_z = 4.0/300 = 0.0133$.

In an ideal elastic material Hooke's law relates the axial strains to the axial stresses by the following linear equations:

$$\varepsilon_x = \frac{1}{E_s}(\sigma_x - \mu\sigma_y - \mu\sigma_z),$$

$$\varepsilon_y = \frac{1}{E_s}(\sigma_y - \mu\sigma_z - \mu\sigma_x),$$

$$\varepsilon_z = \frac{1}{E_s}(\sigma_z - \mu\sigma_x - \mu\sigma_y) \tag{5.1}$$

In axisymmetric conditions (about the $z$-axis) $\sigma_x = \sigma_y$, $\varepsilon_x = \varepsilon_y$. Furthermore there is no lateral strain in the sample of this example and therefore: $\varepsilon_x = \varepsilon_y = 0$. Enforcing these conditions on the above first or second equation we obtain:

$$\sigma_x = \sigma_y = \frac{\mu}{1-\mu}\sigma_z \tag{5.2}$$

Therefore: $\sigma_x = \sigma_y = \dfrac{0.2}{1-0.2}\times\sigma_z = 0.25\sigma_z = 0.25\times300.0 = 75.0\,\text{kPa}$.

$$\varepsilon_z = 0.0133 = \frac{1}{E_s}(\sigma_z - \mu\sigma_x - \mu\sigma_y) = \frac{1}{E_s}(300.0 - 0.2\times75.0 - 0.2\times75.0),$$

$$0.0133 = \frac{270.0}{E_s} \rightarrow E_s \approx 20300\,\text{kPa}.$$

Problem 5.2

Referring to Figure 5.1(b), calculate the distribution of vertical stress along a vertical plane passing through one of the two forces. Specify the values of stress at depths of 1 m, 2 m, 3 m, 4 m and 5 m.

Solution:

With the geometry defined in Figure 5.1(a) the following equations represent the vertical stress component along the depth $z$:

Figure 5.1. Problem 5.2.

$$\sigma_z = \frac{Q}{2\pi z^2}(3\cos^5\theta), \text{ or} \tag{5.3}$$

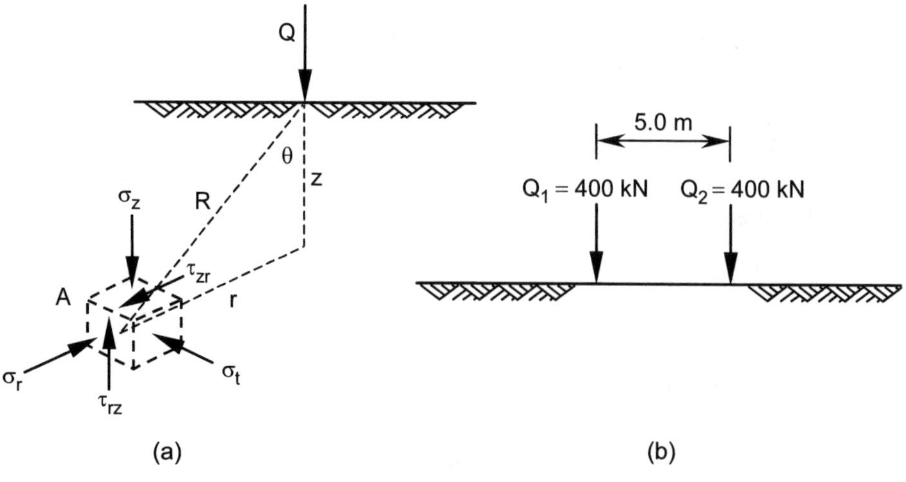

(a)                                                (b)

$$\sigma_z = \frac{Q}{2\pi} \frac{3z^3}{(r^2+z^2)^{5/2}}, \text{ or} \qquad (5.4)$$

$$\sigma_z = \frac{Q}{z^2} \frac{3}{2\pi[(r/z)^2+1]^{5/2}} = \frac{Q}{z^2} I_q,$$

$$I_q = \frac{3}{2\pi[(r/z)^2+1]^{5/2}} \qquad (5.5)$$

At $z = 1.0$ m and under the load 1(using Equation 5.4):

$$\sigma_{z1}(\text{due to } Q_1) = \frac{400.0}{2\pi} \frac{3(1.0)^3}{(0.0^2+1.0^2)^{5/2}} = 190.99 \text{ kPa}.$$

$$\sigma_{z2}(\text{due to } Q_2) = \frac{400.0}{2\pi} \frac{3(1.0)^3}{(5.0^2+1.0^2)^{5/2}} = 0.06 \text{ kPa}.$$

$$\sigma_z = \sigma_{z1} + \sigma_{z2} = 190.99 + 0.06 \approx 191.0 \text{ kPa}.$$

Similar calculations for depths 2 m, 3 m, 4 m and 5 m are carried out and tabulated below.

| Depth (m) | 1.0 | 2.0 | 3.0 | 4.0 | 5.0 |
|-----------|-----|-----|-----|-----|-----|
| $\sigma_z$ (kPa) | 191.0 | 48.1 | 22.0 | 13.1 | 9.0 |

**Problem 5.3**

Figure 5.2(a) shows a plan view of a footing that applies a uniform load of 300 kPa on a horizontal ground surface. Calculate the vertical stress component at point $A$ at a depth of 2 m using:
(a) the principle of superposition of vertical forces by dividing the area into elemental 1 m squares (Figure 5.2(b)),
(b) Newmark's influence chart.

| Load no. | $r$ (m) | $\sigma_z$ (kPa) | Load no. | $r$ (m) | $\sigma_z$ (kPa) | Load no. | $r$ (m) | $\sigma_z$ (kPa) |
|------|-------|------|------|-------|-------|------|-------|-------|
| 1 | 3.808 | 0.78 | 10 | 1.581 | 10.64 | 19 | 2.121 | 5.44 |
| 2 | 2.915 | 2.07 | 11 | 0.707 | 26.68 | 20 | 2.915 | 2.07 |
| 3 | 2.121 | 5.44 | 12 | 0.707 | 26.68 | 21 | 2.549 | 3.21 |
| 4 | 1.581 | 10.64 | 13 | 1.581 | 10.64 | 22 | 2.915 | 2.07 |
| 5 | 1.581 | 10.64 | 14 | 2.549 | 3.21 | 23 | 3.535 | 1.04 |
| 6 | 2.121 | 5.44 | 15 | 0.707 | 26.68 | 24 | 3.535 | 1.04 |
| 7 | 2.915 | 2.07 | 16 | 1.581 | 10.64 | 25 | 3.808 | 0.78 |
| 8 | 3.535 | 1.04 | 17 | 2.549 | 3.21 | 26 | 4.301 | 0.48 |
| 9 | 2.549 | 3.21 | 18 | 1.581 | 10.64 | | Total: | 186.5 |

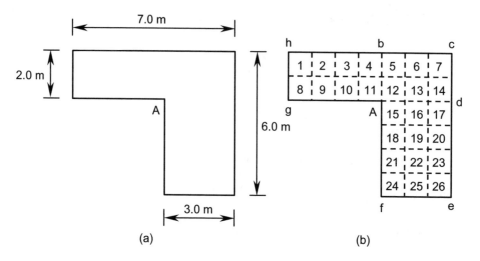

Figure 5.2. Problem 5.3: (a) plan view of a footing, (b) dividing the area into elemental 1 m squares.

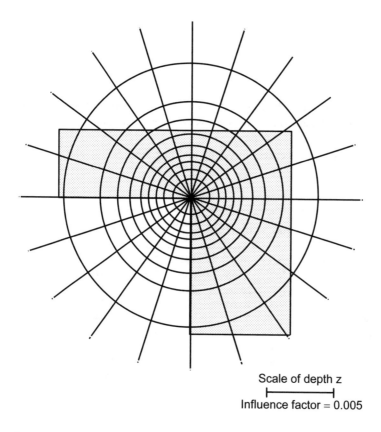

Scale of depth z

Influence factor = 0.005

Figure 5.3. Newmark's influence chart used in the solution of Problem 5.3.

Solution:

(a) The results of computations (using Equation 5.4) are tabulated above where the total vertical stress component under point $A$ at the depth 2 m is 186.5 kPa.

(b) A plan is drawn by assuming that the depth under point $A$ is equal to the length of the scale in Figure 5.3. The re-scaled figure is adjusted on the Newmark's chart so that the point (under which the vertical stress component is required) is located at the central point of the chart. The vertical stress component due to a unit contact pressure is equal to the number of elements within the plan multiplied by the influence factor, or

$$\sigma_z = (\text{number of elements covered}) \times I_q \times q \tag{5.6}$$

where $I_q = 0.005$ (for the chart of Figure 5.3) is called the influence factor.

The best estimate for the number of elements is 123. Therefore:

$$\sigma_z = 123 \times 0.005 \times 300.0 = 184.5 \text{ kPa}.$$

Problem 5.4

Re-work Problem 5.3 using Fadum's equations (or chart).

Solution:

The vertical stress component at a specified depth under a corner of a rectangular loaded area is calculated from:

$$\sigma_z = q \times I_q \tag{5.7}$$

where $I_q$ is the influence factor, a dimensionless parameter defined by:

$$I_q = \frac{1}{4\pi} \left( \frac{2mn\sqrt{m^2 + n^2 + 1}}{m^2 + n^2 + 1 + m^2 n^2} \frac{m^2 + n^2 + 2}{m^2 + n^2 + 1} + \tan^{-1} \frac{2mn\sqrt{m^2 + n^2 + 1}}{m^2 + n^2 + 1 - m^2 n^2} \right) \tag{5.8}$$

and $m$ and $n$ are interchangeable parameters defined as:

$$m = \frac{L}{z}, \quad n = \frac{B}{z} \tag{5.9}$$

where $L$ and $B$ are the plan dimensions of the loaded rectangle. For negative values of $\tan^{-1}$, $\pi$ should be added to the calculated angle measured in radians.

The loaded area is divided into the 3 rectangles *Abcd*, *Adef*, *Aghb* as shown in Figure 5.2(b). Note that all rectangles share a common corner of $A$.

Results are tabulated below.

| Rectangle | $L$ (m) | $B$ (m) | $m$ | $n$ | $I_q$ | $\sigma_z$ (kPa) |
|-----------|---------|---------|-----|-----|-------|------------------|
| *Abcd* | 3.0 | 2.0 | 1.5 | 1.0 | 0.1936 | 58.08 |
| *Adef* | 4.0 | 3.0 | 2.0 | 1.5 | 0.2236 | 67.08 |
| *Aghb* | 4.0 | 2.0 | 2.0 | 1.0 | 0.1999 | 59.97 |
| | | | Total vertical stress component at $A = 185.1$ kPa. | | | |

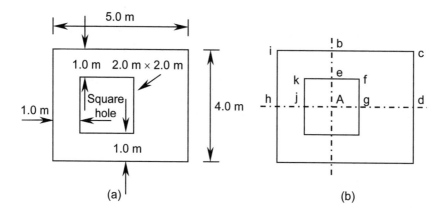

Figure 5.4. Problem 5.5.

Problem 5.5

Figure 5.4(a) shows a plan view of a rectangular footing with a 2 m × 2 m square hole (through its entire thickness). The hole is located at 1 m from the left edge and is equally positioned between the top and lower edges of the footing. If the uniform contact pressure under the footing is 200 kPa, use Fadum's equation (or chart) to compute the vertical stress component at a point 2 m below the centre of the square hole.

Solution:

The centre of the square hole is outside the loaded area then a few of the rectangles that share this common corner cover areas that are not loaded. In this case it is convenient to assume that the rectangles that cover the unloaded areas are subjected to a negative contact pressure, e.g.:

vertical stress at depth $z$ due to area *ebcdgf* (Figure 5.4(b)) =
vertical stress due to area *abcd* − vertical stress due to area *Aefg*.

The results of the computations are summarized in the table below. Because of symmetry only the upper half of the footing is considered.

| Rectangle | $L$ (m) | $B$ (m) | $m$ | $n$ | $I_q$ | $\sigma_z$ (kPa) |
|---|---|---|---|---|---|---|
| *Abcd* | 3.0 | 2.0 | 1.5 | 1.0 | 0.1936 | 38.72 |
| *Ahib* | 2.0 | 2.0 | 1.0 | 1.0 | 0.1752 | 35.04 |
| *Aefg* | 1.0 | 1.0 | 0.5 | 0.5 | 0.0840 | −16.80 |
| *Ajke* | 1.0 | 1.0 | 0.5 | 0.5 | 0.0840 | −16.80 |
| | | | | | Total: | 40.16 |
| Total vertical stress component at $A = 2 \times 40.16 = 80.3$ kPa. | | | | | | |

Problem 5.6

Compute $\sigma_x$, $\sigma_z$ and $\tau_{xz}$ at point $A$ (Figure 5.5(b)) due to the two line loads shown.

Solution:

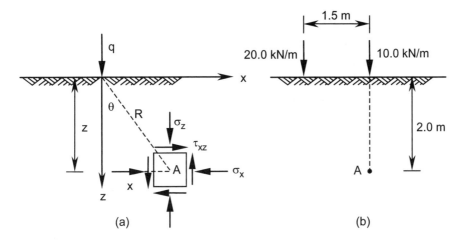

Figure 5.5. Problem 5.6.

The stress components at depth $z$ are calculated from the following equations with sign conventions shown in Figure 5.5(a).

$$\sigma_x = \frac{2q}{\pi z}\cos^2\theta\sin^2\theta = \frac{2qx^2 z}{\pi(x^2 + z^2)^2} \tag{5.10}$$

$$\sigma_z = \frac{2q}{\pi z}\cos^4\theta = \frac{2qz^3}{\pi(x^2 + z^2)^2} \tag{5.11}$$

$$\tau_{xz} = \frac{2q}{\pi z}\cos^3\theta\sin\theta = \frac{2qxz^2}{\pi(x^2 + z^2)^2} \tag{5.12}$$

For the line load of 10 kN/m the coordinates of point $A$ are:
$x_A = 0.0$, $z_A = 2.0$ m:

$\sigma_{x1} = 0.0$.

$$\sigma_{z1} = \frac{2qz^3}{\pi(x^2 + z^2)^2} = \frac{2\times10.0\times2.0^3}{\pi(0.0^2 + 2.0^2)^2} = 3.18 \text{ kPa.}$$

$\tau_{xz1} = 0.0$.

For the line load of 20 kN/m the coordinates of point $A$ are:
$x_A = 1.5$ m, $z_A = 2.0$ m:

$$\sigma_{x2} = \frac{2qx^2 z}{\pi(x^2 + z^2)^2} = \frac{2\times20.0\times1.5^2\times2.0}{\pi(1.5^2 + 2.0^2)^2} = 1.5 \text{ kPa.}$$

$$\sigma_{z2} = \frac{2qz^3}{\pi(x^2 + z^2)^2} = \frac{2\times20.0\times2.0^3}{\pi(1.5^2 + 2.0^2)^2} = 2.61 \text{ kPa.}$$

$$\tau_{xz2} = \frac{2qxz^2}{\pi(x^2+z^2)^2} = \frac{2\times20.0\times1.5\times2.0^2}{\pi(1.5^2+2.0^2)^2} = 2.0\,\text{kPa}.$$

Superposing two results:

$$\sigma_x = \sigma_{x1} + \sigma_{x2} = 0.0+1.5 = 1.5\,\text{kPa}.$$
$$\sigma_z = \sigma_{z1} + \sigma_{z2} = 3.18+2.61 = 5.8\,\text{kPa}.$$
$$\tau_z = \tau_{xz1} + \tau_{xz2} = 0.0+2.0 = 2.0\,\text{kPa}.$$

**Problem 5.7**

Compute the magnitudes of the major and minor principal stresses within a soil that is subjected to an infinitely long line load applied at the ground surface.

Solution:

Using Equations 5.10 to 5.12 the position of the centre and the radius of the Mohr's circle of stress are obtained by:

$$s = \frac{\sigma_z+\sigma_x}{2} = \frac{q}{\pi z}(\cos^4\theta+\cos^2\theta\sin^2\theta),$$

$$s = \frac{q}{\pi z}\cos^2\theta.$$

$$t = \sqrt{[(\sigma_z-\sigma_x)/2]^2+\tau_{zx}^2} = \sqrt{\left[\frac{q}{\pi z}(\cos^4\theta-\cos^2\theta\sin^2\theta)\right]^2+(\frac{2q}{\pi z}\cos^3\theta\sin\theta)^2},$$

$$t = \frac{q\cos^2\theta}{\pi z}\sqrt{[(\cos^2\theta-\sin^2\theta)]^2+(2\cos\theta\sin\theta)^2},$$

$$t = \frac{q\cos^2\theta}{\pi z}\sqrt{\cos^2 2\theta+\sin^2 2\theta} = \frac{q}{\pi z}\cos^2\theta.$$

$$\sigma_3 = s-t = 0.$$

$$\sigma_1 = s+t = \frac{q}{\pi z}\cos^2\theta+\frac{q}{\pi z}\cos^2\theta = \frac{2q}{\pi z}\cos^2\theta.$$

The direction of the major principal stress plane is calculated from:

$$\tan 2\theta_1 = \frac{2\tau_{xz}}{\sigma_z-\sigma_x} = \frac{2\frac{2q}{\pi z}\cos^3\theta\sin\theta}{\frac{2q}{\pi z}(\cos^4\theta-\cos^2\theta\sin^2\theta)},$$

$$\tan 2\theta_1 = \frac{2\cos^3\theta\sin\theta}{\cos^4\theta-\cos^2\theta\sin^2\theta} = \frac{2\cos\theta\sin\theta}{\cos^2\theta-\sin^2\theta} = \frac{\sin 2\theta}{\cos 2\theta} = \tan 2\theta.$$

This means that at any point the angle between the major principal plane and the *x*-axis is θ or the major principal stress is on the line connecting the point of interest to the application point of the line load.

Problem 5.8

Under the centre line of an infinite strip footing, calculate the depth at which the vertical stress component is 10% of the applied load.

Solution:

Referring to Figure 5.6 the vertical stress component is according:

$$\sigma_z = \frac{q}{\pi}\left[\alpha + \sin \alpha \cos(\alpha + 2\beta)\right] \qquad (5.13)$$

Angles $\alpha$ and $\beta$ are defined by:

$$\alpha = \tan^{-1}\frac{x+b}{z} - \tan^{-1}\frac{x-b}{z}, \quad \beta = \tan^{-1}\frac{x-b}{z} \qquad (5.14)$$

Along the centre line $x = 0$, thus

$$\alpha = \tan^{-1}\frac{b}{z} - \tan^{-1}\frac{-b}{z} = 2\tan^{-1}\frac{b}{z}, \quad \beta = \tan^{-1}\frac{-b}{z},$$

which means $\beta = -\alpha/2$. Substituting these conditions in Equation 5.13 and setting the result to $0.1q$:

$$\sigma_z = 0.1q = \frac{q}{\pi}\left[\alpha + \sin \alpha \cos(\alpha - \alpha)\right],$$

$0.1\pi = \alpha + \sin \alpha$.

A trial and error method yields $\alpha = 9.02°$.

$$\alpha = 9.02° = 2\tan^{-1}\frac{b}{z},$$

$$\frac{z}{b} = 12.68 \rightarrow \frac{z}{2b} = \frac{z}{B} = 6.34.$$

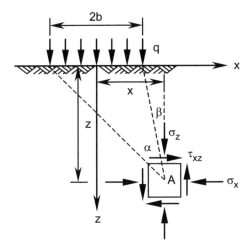

Figure 5.6. Problem 5.8.

Problem 5.9

A strip loading of infinite length is shown in Figure 5.7. Compute the vertical stress component at points $A$, $B$, $C$, and $D$.

Solution:

At point $A$, $x = 0$, $z = 1.0$ m:
$$\alpha = \tan^{-1}[(0.0+1.0)/1.0] - \tan^{-1}[(0.0-1.0)/1.0] = 90.0°,$$
$$\beta = \tan^{-1}[(0.0-1.0)/1.0] = -45.0°.$$

Using Equation 5.13:
$$\sigma_z = \frac{100.0}{\pi}\left[\frac{90.0°\times\pi}{180.0°} + \sin 90.0°\cos(90.0° - 2\times 45.0°)\right] = 81.8 \text{ kPa.}$$

At point $B$, $x = 1.0$ m, $z = 1.0$ m:
$$\alpha = \tan^{-1}[(1.0+1.0)/1.0] - \tan^{-1}[(1.0-1.0)/1.0] = 63.43°,$$
$$\beta = \tan^{-1}[(1.0-1.0)/1.0] = 0.0°.$$

$$\sigma_z = \frac{100.0}{\pi}\left[\frac{63.43°\times\pi}{180.0°} + \sin 63.43°\cos(63.43° + 2\times 0.0)\right] = 48.0 \text{ kPa.}$$

At point $C$, $x = 2.0$, $z = 1.0$ m:
$$\alpha = \tan^{-1}[(2.0+1.0)/1.0] - \tan^{-1}[(2.0-1.0)/1.0] = 26.56°,$$
$$\beta = \tan^{-1}[(2.0-1.0)/1.0] = 45.0°.$$

$$\sigma_z = \frac{100.0}{\pi}\left[\frac{26.56°\times\pi}{180.0°} + \sin 26.56°\cos(26.56° + 2\times 45.0°)\right] = 8.4 \text{ kPa.}$$

At point $D$, $x = 3.0$, $z = 1.0$ m:
$$\alpha = \tan^{-1}[(3.0+1.0)/1.0] - \tan^{-1}[(3.0-1.0)/1.0] = 12.53°,$$
$$\beta = \tan^{-1}[(3.0-1.0)/1.0] = 64.43°.$$

$$\sigma_z = \frac{100.0}{\pi}\left[\frac{12.53°\times\pi}{180.0°} + \sin 12.53°\cos(12.53° + 2\times 64.43°)\right] = 1.7 \text{ kPa.}$$

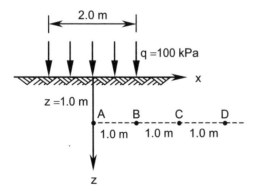

Figure 5.7. Problem 5.9.

Problem 5.10

An earth embankment is 2.5 m high and has a slope of 2 horizontal to 1 vertical. If the base of the embankment is 20 m, find the vertical stress component at a point on the centre line at a depth of 7 m (measured from the base). The unit weight of the embankment material is 18 kN/m³.

Solution:

Figure 5.8(a) shows a vertical distributed load applied to an infinitely long strip and linearly increasing across its width 2b. Using the sign convention shown, the stress components are:

$$\sigma_x = \frac{q}{2\pi}\left(\frac{x}{b}\alpha - \frac{z}{b}\ln\frac{R_1^2}{R_2^2} + \sin 2\beta\right) \tag{5.15}$$

$$\sigma_z = \frac{q}{2\pi}\left(\frac{x}{b}\alpha - \sin 2\beta\right) \tag{5.16}$$

$$\tau_{xz} = \frac{q}{2\pi}\left(1 + \cos 2\beta - \frac{z}{b}\alpha\right) \tag{5.17}$$

Angles $\alpha$ and $\beta$ are defined by:

$$\alpha = \tan^{-1}\frac{x}{z} - \tan^{-1}\frac{x-2b}{z}, \ \beta = \tan^{-1}\frac{x-2b}{z} \tag{5.18}$$

The vertical stress component under a linear and uniform loading combination shown in Figure 5.8(b) is obtained by combining Equations 5.13 and 5.16:

$$\sigma_z = \frac{q}{\pi}\left(\alpha_2 + \alpha_1\frac{a+b}{a}\right) \tag{5.19}$$

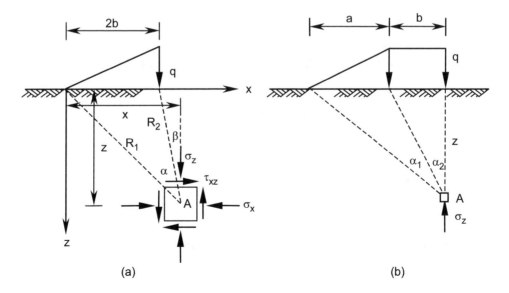

(a)                                                              (b)

Figure 5.8. Problem 5.10.

where $\alpha_1$ and $\alpha_2$ are calculated from Equations 5.18:

$$\alpha_1 = \tan^{-1}\frac{a+b}{z} - \alpha_2, \ \alpha_2 = \tan^{-1}\frac{b}{z} \tag{5.20}$$

Referring to Figure 5.8(b):
$a = 2.5$ (the height of embankment) $\times 2 = 5.0$ m.
$b$ = half of the width of the embankment at the top,
$b = (20.0 - 2 \times a)/2 = (20.0 - 2 \times 5.0)/2 = 5.0$ m. From Equations 5.20:
$\alpha_2 = \tan^{-1}(5.0/7.0) = 35.54°$,
$\alpha_1 = \tan^{-1}[(5.0+5.0)/7.0] - \alpha_2 = 55.00° - 35.54° = 19.46°$.
$q = 18.0 \times 2.5 = 45.0$ kPa.
From Equation 5.19 and multiplying the result by 2:

$$\sigma_z = \frac{2 \times 45.0}{\pi}\left(\frac{35.54° \times \pi}{180.0°} + \frac{19.46° \times \pi}{180.0°} \times \frac{5.0+5.0}{5.0}\right) = 37.2 \text{ kPa.}$$

## Problem 5.11

A vertical pile carrying a load of 1500 kN has been driven 18 m into the ground. Calculate the vertical stress component at a point 19.8 m below the ground surface and 3 m from the centre line of the pile for the following cases:
(a) the entire load is transmitted to the soil through the base of the pile,
(b) the base carries one half of the load and the rest is carried by skin friction. $\mu = 0.3$.

Solution:

The vertical stress component at any point with a horizontal distance $r$ from the load $Q$ and a depth $z > L$ is:

$$\sigma_z = \frac{Q}{L^2}I_q \tag{5.21}$$

where $L$ is the depth of vertical load $Q$ and $I_q$ is given by Tables 5.1(a) and 5.1(b) (for $\mu = 0.3$). The coordinates of the point of interest are expressed in dimensionless forms of $m = z/L$, and $n = r/L$.
$m = z/L = 19.8/18.0 = 1.1$, and $n = r/L = 3.0/18.0 = 0.167$.
(a) Using Table 5.1(a) and interpolating $I_q$ values between $r = 0.1$ and $r = 0.2$:

$$I_q = 3.9054 - \frac{(3.9054 - 0.5978)}{0.1} \times (0.167 - 0.1) = 1.7003.$$

$$\sigma_z = \frac{Q}{L^2}I_q = \frac{1500.0}{18.0^2} \times 1.7003 = 7.9 \text{ kPa.}$$

(b) From Table 5.1(b) and interpolating $I_q$ values between $r = 0.15$ and $r = 0.2$:

$$I_q = 0.8368 - \frac{(0.8368 - 0.6419)}{0.05} \times (0.167 - 0.15) = 0.7705.$$

$$\sigma_z = 7.9/2 + \frac{(1500.0/2)}{18.0^2} \times 0.7705 = 5.7 \text{ kPa.}$$

Table 5.1(a). Vertical stress influence factors due to a vertical load applied in the interior of the soil ($\mu = 0.3$).

| $n \rightarrow$ $m \downarrow$ | 0.00 | 0.1 | 0.20 | 0.30 | 0.40 | 0.50 | 0.75 | 1.00 | 1.50 | 2.00 |
|---|---|---|---|---|---|---|---|---|---|---|
| 1.0 | $\infty$ | 0.1013 | 0.0986 | 0.0944 | 0.0889 | 0.0824 | 0.0641 | 0.0463 | 0.0209 | 0.0087 |
| 1.1 | 19.3926 | 3.9054 | 0.5978 | 0.2123 | 0.1287 | 0.0986 | 0.0668 | 0.0475 | 0.0222 | 0.0097 |
| 1.2 | 4.9099 | 2.9275 | 1.0358 | 0.4001 | 0.2027 | 0.1303 | 0.0722 | 0.0493 | 0.0235 | 0.0106 |
| 1.3 | 2.2222 | 1.7467 | 0.9757 | 0.4970 | 0.2717 | 0.1687 | 0.0808 | 0.0519 | 0.0247 | 0.0116 |
| 1.4 | 1.2777 | 1.1152 | 0.7905 | 0.4891 | 0.3032 | 0.1974 | 0.0908 | 0.0555 | 0.0260 | 0.0125 |
| 1.5 | 0.8377 | 0.7686 | 0.6070 | 0.4356 | 0.3012 | 0.2098 | 0.0999 | 0.0594 | 0.0274 | 0.0134 |
| 1.6 | 0.5968 | 0.5626 | 0.4768 | 0.3738 | 0.2809 | 0.2086 | 0.1063 | 0.0631 | 0.0288 | 0.0143 |
| 1.7 | 0.4500 | 0.4312 | 0.3819 | 0.3177 | 0.2538 | 0.1988 | 0.1094 | 0.0661 | 0.0302 | 0.0152 |
| 1.8 | 0.3536 | 0.3424 | 0.3122 | 0.2706 | 0.2262 | 0.1849 | 0.1096 | 0.0682 | 0.0315 | 0.0161 |
| 1.9 | 0.2866 | 0.2795 | 0.2600 | 0.2321 | 0.2006 | 0.1697 | 0.1076 | 0.0693 | 0.0326 | 0.0169 |
| 2.0 | 0.2380 | 0.2333 | 0.2201 | 0.2007 | 0.1780 | 0.1547 | 0.1039 | 0.0694 | 0.0336 | 0.0177 |

Table 5.1(b). Vertical stress influence factors due to uniform shear force applied in the interior of the soil ($\mu = 0.3$).

| $n \rightarrow$ $m \downarrow$ | 0.00 | 0.02 | 0.04 | 0.06 | 0.08 | 0.10 | 0.15 | 0.20 | 0.50 | 1.00 | 2.00 |
|---|---|---|---|---|---|---|---|---|---|---|---|
| 1.0 | $\infty$ | 6.8419 | 3.4044 | 2.2673 | 1.6983 | 1.3567 | 0.8998 | 0.6695 | 0.2346 | 0.0686 | 0.0076 |
| 1.1 | 1.9219 | 1.8611 | 1.7072 | 1.5134 | 1.3211 | 1.1503 | 0.8368 | 0.6419 | 0.2335 | 0.0728 | 0.0091 |
| 1.2 | 0.9699 | 0.9403 | 0.9166 | 0.8825 | 0.8400 | 0.7922 | 0.6688 | 0.5588 | 0.2292 | 0.0760 | 0.0105 |
| 1.3 | 0.6430 | 0.6188 | 0.6099 | 0.5992 | 0.5850 | 0.5675 | 0.5157 | 0.4597 | 0.2207 | 0.0782 | 0.0120 |
| 1.4 | 0.4867 | 0.4558 | 0.4507 | 0.4461 | 0.4396 | 0.4316 | 0.4063 | 0.3761 | 0.2082 | 0.0796 | 0.0134 |
| 1.5 | 0.3766 | 0.3561 | 0.3533 | 0.3510 | 0.3476 | 0.3432 | 0.3291 | 0.3115 | 0.1834 | 0.0800 | 0.0148 |
| 1.6 | 0.3339 | 0.2895 | 0.2878 | 0.2863 | 0.2843 | 0.2817 | 0.2732 | 0.2621 | 0.1777 | 0.0796 | 0.0160 |
| 1.7 | 0.2664 | 0.2438 | 0.2414 | 0.2399 | 0.2384 | 0.2369 | 0.2313 | 0.2239 | 0.1623 | 0.0784 | 0.0172 |
| 1.8 | 0.2025 | 0.2065 | 0.2054 | 0.2044 | 0.2038 | 0.2026 | 0.1989 | 0.1938 | 0.1479 | 0.0766 | 0.0182 |
| 1.9 | 0.1847 | 0.1794 | 0.1785 | 0.1777 | 0.1768 | 0.1760 | 0.1733 | 0.1696 | 0.1347 | 0.0744 | 0.0191 |
| 2.0 | 0.1634 | 0.1565 | 0.1561 | 0.1556 | 0.1551 | 0.1545 | 0.1525 | 0.1498 | 0.1229 | 0.0718 | 0.0199 |

## Problem 5.12

Two flexible square footings 2 m × 2 m are constructed 4 m apart (centre to centre) on the ground surface (Figure 5.9). The uniform contact pressure under one footing is 200 kPa and is 400 kPa under the other. Calculate the elastic settlement at the centre of each footing and at the mid-point of the line connecting the two centres. $E_s = 10000$ kPa, $\mu = 0.35$.

Solution:

The elastic settlement $S_e$ of a flexible footing, either rectangular of dimensions $L \times B$ ($L > B$) or circular of diameter $B$, is given by:

$$S_e = qB\frac{1-\mu^2}{E_s}I_s \tag{5.22}$$

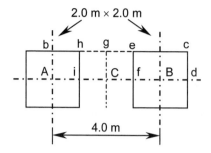

Figure 5.9. Problem 5.12.

where $I_s$ is an influence factor depending on the shape and $L / B$ ratio. The influence factor for a flexible rectangular footing at the corner of a loaded rectangular area is:

$$I_1 = \frac{1}{\pi}\left[\frac{L}{B}\ln\left(\frac{1+\sqrt{(L/B)^2+1}}{L/B}\right)+\ln\left(\frac{L}{B}+\sqrt{(L/B)^2+1}\right)\right] \qquad (5.23)$$

$$I_s\,(\text{centre}) = 2\,I_1 \qquad (5.24)$$

$$I_s\,(\text{average}) = 1.696\,I_1 \qquad (5.25)$$

For a rigid rectangular footing the influence factor is approximately 0.926 of the average influence factor of the flexible footing.
Hence, $I_s$ (rigid) $= 0.926 \times I_s$ (average) $= 0.926 \times 1.696\,I_1$, or:

$$I_s\,(\text{rigid}) = 1.570 I_1 = (\pi/2)I_1 \qquad (5.26)$$

For a square loaded area ($L / B = 1$), $I_1 = 0.561$, thus at the centre $I_s = 1.122$. The settlement of each footing with no contributions from the adjacent loadings is:

$$S_e\,(\text{for }q\ =\ 200.0\,\text{kPa}) = 200.0 \times 2.0 \times \frac{1-0.35^2}{10000.0} \times 1.122 = 0.0394\,\text{m}.$$

$$S_e\,(\text{for }q\ =\ 400.0\,\text{kPa}) = 0.0788\,\text{m}.$$

Contribution of the adjacent loads are calculated and summarized in the second and third rows of the table below:

| Rectangle | $L$ (m) | $B$ (m) | $L/B$ | $I_1$ |
|---|---|---|---|---|
| *Abcd* | 5.0 | 1.0 | 5.0 | 1.052 |
| *Abef* | 3.0 | 1.0 | 3.0 | −0.891 |
| *Cgcd* | 3.0 | 1.0 | 3.0 | 0.891 |
| *Cgef* | 1.0 | 1.0 | 1.0 | −0.561 |

$$S_e\,(\text{for }q\ =\ 200.0\,\text{kPa}) = 0.0394 + 2 \times 400.0 \times 1.0 \times \frac{1-0.35^2}{10000.0}(1.052-0.891) = 0.0507\,\text{m}.$$

$$S_e\,(\text{for }q\ =\ 200.0\,\text{kPa}) \approx 51\,\text{mm}.$$

$$S_e \text{ (for } q = 400.0 \text{ kPa)} = 0.0788 + 2 \times 200.0 \times 1.0 \times \frac{1 - 0.35^2}{10000.0}(1.052 - 0.891) = 0.0844 \text{ m.}$$

$S_e$ (for $q = 400.0$ kPa) $\approx 84.5$ mm.

The areas involved at point $C$ are explained in the lower two rows of the table above. Because of the geometrical symmetry only areas at the right hand of this point are considered; hence:

$$S_e \text{ (at point } C) = 2 \times 200.0 \times 1.0 \times \frac{1 - 0.35^2}{10000.0}(0.891 - 0.561) +$$

$$2 \times 400.0 \times 1.0 \times \frac{1 - 0.35^2}{10000.0}(0.891 - 0.561) = 0.0347 \text{ m, } S_e \text{ (at point } C) \approx 34.7 \text{ mm.}$$

## Problem 5.13

A flexible circular footing 4 m in diameter exerts a pressure 300 kPa at the ground surface. The soil is a saturated clay with $E_s = 8000$ kPa, and an incompressible stratum is located 12 m below the ground surface. Calculate the settlement under the centre of the footing using:
(a) a semi-numerical method by dividing the depth into 6 layers of 2 m thickness,
(b) Steinbrenner's influence factors.

Solution:

(a) For the point that is located on a vertical axis passing through the centre of the loaded circle, the integration process for the vertical stress component (Equation 5.4 or 5.5) is straightforward and yields:

$$\sigma_z = q \left\{ 1 - \frac{1}{[(R_C/z)^2 + 1]^{3/2}} \right\} \tag{5.27}$$

where $R_C$ is the radius of the loaded area. The radial and tangential stress components (Figure 5.1) are:

$$\sigma_r = \sigma_t = \frac{q}{2} \left\{ 1 + 2\mu - \frac{2(1+\mu)}{[(R_C/z)^2 + 1]^{1/2}} + \frac{1}{[(R_C/z)^2 + 1]^{3/2}} \right\} \tag{5.28}$$

Using Equations 5.27 and 5.28 along with Hooke's law (Equations 5.1) the following results are obtained. For the saturated clay we assume $\mu = 0.5$.

| Layer | Depth (m) | $\sigma_z$ (kPa) | $\sigma_r$ (kPa) | $\sigma_z - 2\mu \sigma_r$ (kPa) | $\varepsilon_z \times 10^{-3}$ | $\Delta s$ (mm) |
|---|---|---|---|---|---|---|
| 1 | 1.0 | 273.17 | 112.17 | 161.00 | 20.12 | 40.24 |
| 2 | 3.0 | 127.19 | 11.98 | 115.21 | 14.40 | 28.80 |
| 3 | 5.0 | 59.88 | 2.25 | 57.63 | 7.20 | 14.40 |
| 4 | 7.0 | 33.31 | 0.66 | 32.65 | 4.08 | 8.16 |
| 5 | 9.0 | 20.92 | 0.25 | 20.67 | 2.58 | 5.16 |
| 6 | 11.0 | 14.28 | 0.12 | 14.16 | 1.77 | 3.54 |
| | | | | | Total: | $\approx 100$ |

Sample calculation for layer 3:

$$\sigma_z = 300.0 \left\{ 1 - \frac{1}{[(2.0/5.0)^2 + 1]^{3/2}} \right\} = 59.88 \text{ kPa.}$$

$$\sigma_r = \sigma_t = \frac{150.0}{2} \left\{ 1 + 2 \times 0.5 - \frac{2(1+0.5)}{[(2.0/5.0)^2 + 1]^{1/2}} + \frac{1}{[(2.0/5.0)^2 + 1]^{3/2}} \right\} = 2.25 \text{ kPa.}$$

$$\varepsilon_z = \frac{(\sigma_z - \mu\sigma_r - \mu\sigma_t)}{E_s} = \frac{(\sigma_z - 2\mu\sigma_r)}{8000.0} = \frac{(59.88 - 2 \times 2.25)}{8000.0} = \frac{57.63}{8000.0} = 7.20 \times 10^{-3}.$$

$\Delta s = \varepsilon_z \times z_l = 7.20 \times 10^{-3} \times 2000.0 = 14.4 \text{ mm.}$

(b) If an incompressible layer (for which $E_s \rightarrow \infty$) is located at depth $H$ from the ground surface, the $I_s$ values for a flexible rectangular footing are given by:

$$I_s(\text{corner}) = I_1 = I_2 + \frac{1-2\mu}{1-\mu} I_3, \; I_s(\text{centre}) = 2I_1, \; I_s(\text{average}) = 1.696 I_1 \quad (5.29)$$

$$I_2 = \frac{1}{\pi} \left\{ \frac{L}{B} \ln \left[ \frac{\left(1 + \sqrt{(L/B)^2 + 1}\right)\sqrt{(L/B)^2 + (H/B)^2}}{L/B\left(1 + \sqrt{(L/B)^2 + (H/B)^2 + 1}\right)} \right] + \right.$$

$$\left. \ln \left[ \frac{\left(L/B + \sqrt{(L/B)^2 + 1}\right)\sqrt{1 + (H/B)^2}}{L/B + \sqrt{(L/B)^2 + (H/B)^2 + 1}} \right] \right\} \quad (5.30)$$

$$I_3 = \frac{(H/B)}{2\pi} \tan^{-1} \left[ \frac{L/B}{(H/B)\sqrt{(L/B)^2 + (H/B)^2 + 1}} \right] \quad (5.31)$$

The equivalent $L$ or $B$ is: $\sqrt{\pi D^2 / 4} = \sqrt{\pi 4.0^2 / 4} = 3.545 \text{ m.}$

Thus $I_2 = 0.383$, $I_3 = 0.0432$, $I_1 = 0.383$, $I_s(\text{centre}) = 2I_1 = 2 \times 0.383 = 0.766$.

$$S_e(\text{at centre}) = 300.0 \times 3.545 \times \frac{1 - 0.5^2}{8000.0} \times 0.766 = 76.4 \approx 77 \text{ mm.}$$

Problem 5.14

A flexible rectangular footing for which $B = 2$ m and $L = 4$ m, exerts 300 kPa, 2 m below the ground surface. The Modulus of Elasticity for the soil is 8000 kPa, and $\mu = 0.5$. An incompressible stratum is located at 10 m below the ground surface. Calculate the average settlement of the footing using:
(a) Steinbrenner's influence factors and Fox's correction factor for depth,
(b) the improved chart of Janbu et al. (Figure 5.10).
Solution:

(a) Fox's correction factor $I_F$ for the effect of the embedment of a footing ($D$) on the average elastic settlement of a uniformly loaded rectangular area is given in Table 5.2.

Table 5.2. Depth factors $I_F$ ($\mu = 0.5$).

| $D/B$ | $L/B = 1.0$ | $L/B = 2.0$ | $L/B = 3.0$ | $L/B = 4.0$ | $L/B = 5.0$ | $L/B = 10.0$ |
|---|---|---|---|---|---|---|
| 0.00 | 1.0000 | 1.0000 | 1.0000 | 1.0000 | 1.0000 | 1.0000 |
| 0.25 | 0.9424 | 0.9650 | 0.9723 | 0.9760 | 0.9783 | 0.9883 |
| 0.50 | 0.8501 | 0.8993 | 0.9184 | 0.9284 | 0.9348 | 0.9489 |
| 0.75 | 0.7760 | 0.8372 | 0.8651 | 0.8808 | 0.8909 | 0.9139 |
| 1.00 | 0.7234 | 0.7864 | 0.8196 | 0.8394 | 0.8525 | 0.8829 |
| 1.50 | 0.6587 | 0.7150 | 0.7510 | 0.7750 | 0.7919 | 0.8334 |
| 2 | 0.6220 | 0.6698 | 0.7039 | 0.7288 | 0.7473 | 0.7957 |
| 3 | 0.5828 | 0.6180 | 0.6459 | 0.6684 | 0.6868 | 0.7412 |
| 4 | 0.5625 | 0.5899 | 0.6126 | 0.6319 | 0.6485 | 0.7030 |
| 5 | 0.5502 | 0.5724 | 0.5913 | 0.6079 | 0.6226 | 0.6746 |
| 6 | 0.5419 | 0.5606 | 0.5767 | 0.5911 | 0.6040 | 0.6526 |
| 8 | 0.5315 | 0.5456 | 0.5580 | 0.5692 | 0.5796 | 0.6210 |
| 10 | 0.5252 | 0.5366 | 0.5466 | 0.5557 | 0.5643 | 0.5998 |
| 50 | 0.5050 | 0.5073 | 0.5094 | 0.5113 | 0.5131 | 0.5212 |
| 100 | 0.5025 | 0.5037 | 0.5047 | 0.5056 | 0.5065 | 0.5106 |

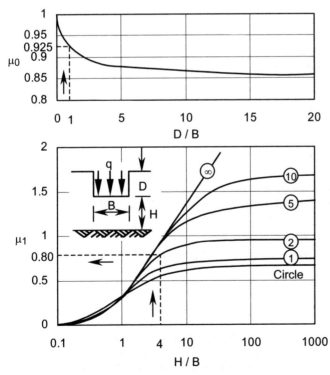

Figure 5.10. Improved (originally Janbu et al., 1956) influence factors $\mu_0$ and $\mu_1$ for saturated clays (Christian & Carrier, 1978) used in the solution of Problem 5.14.

Assume the base of the footing is located at the ground surface, therefore:
$H/B = 10.0/2.0 = 5.0, L/B = 4.0/2.0 = 2.0$.

From Equations 5.30 and 5.31: $I_2 = 0.526, I_3 = 0.058$. Thus using Equation 5.29:
$I_s(\text{corner}) = I_1 = 0.526, I_s(\text{average}) = 1.696 \times 0.526 = 0.892$.

$$S_e(\text{average}) = 300.0 \times 2.0 \times \frac{1 - 0.5^2}{8000.0} \times 0.892 = 0.050 \text{ m} = 50 \text{ mm}.$$

Using Table 5.2 the Fox's depth factor is found to be 0.7864; thus
$S_e(\text{average}) = 50.0 \times 0.7864 = 39.3$ mm.

(b) For the saturated clay the settlement is calculated from:

$$S_e = \mu_0\mu_1\frac{qB}{E_s} \tag{5.32}$$

where $\mu_0$ and $\mu_1$ are influence factors for depth and layer thickness respectively. With definitions described in Figure 5.10:
$H/B = (10.0 - 2.0)/2.0 = 4.0, L/B = 4.0/2.0 = 2.0, D/B = 2.0/2.0 = 1.0$.
From Figure 5.10: $\mu_0 = 0.925$, and $\mu_1 = 0.80$, therefore using Equation 5.32:
$S_e = 0.925 \times 0.80 \times 300.0 \times 2.0/8000.0 = 0.055$ m $= 55$ mm.

## Problem 5.15

A footing has plan dimensions of 3 m × 5 m and exerts a uniform net contact pressure of 200 kPa on the underlying saturated sand. The base of the footing is at the ground surface, and the average *SPT* number at a depth of 3 m is 19. Calculate the settlement of the footing using Burland and Burbidge method.

Solution:

For a normally consolidated sand and square footing the general equation for the settlement is:

$$S_e = B^{0.75}\frac{1.7}{\overline{N}^{1.4}}q \tag{5.33}$$

where $S_e$ is in mm, $q$ is the contact pressure in kPa, $B$ is the width of the footing in m and $\overline{N}$ is the average *SPT* number within the influence depth $Z_I$ defined by:

$$Z_I = B^{0.75} \tag{5.34}$$

The value of $\overline{N}$ is corrected for saturated sands and gravelly sands according:

$$N' = 15 + 0.5(\overline{N} - 15) \qquad \text{saturated sand} \tag{5.35}$$

$$N' = 1.25\overline{N} \qquad \text{gravelly sand} \tag{5.36}$$

For overconsolidated soils, and for $q > p'_c$ (preconsolidation pressure) the settlement is the sum of the recompression and normally consolidated stages, and a different value for compression index: $m_v = 1.7/\overline{N}^{1.4}$ at the recompression side has to be evaluated. Assuming the compressibility index at the recompression stage is 1 / 3 of the normally consolidated stage:

$$S_e = B^{0.75} \frac{1.7}{\overline{N}^{1.4}} (q - 2p'_c/3) \quad \text{for } q > p'_c \tag{5.37}$$

If the applied contact pressure is less than $p'_c$ this becomes:

$$S_e = B^{0.75} \frac{1.7}{\overline{N}^{1.4}} (q/3) \qquad \text{for } q < p'_c \tag{5.38}$$

The above equations are also applicable to footings located at some depth below the ground surface if the preconstruction vertical stress is substituted for $p'_c$. For rectangular footings with $L/B > 1.0$, the settlement is increased by a shape factor according:

$$I_s = \left( \frac{1.25 L/B}{L/B + 0.25} \right)^2 \tag{5.39}$$

If the incompressible layer is located within the influence depth $Z_I$, the calculated settlement is multiplied by a correction factor defined by:

$$I_f = \frac{H(2 - H/Z_I)}{Z_I} \tag{5.40}$$

where $H$ is the depth of the incompressible layer from the footing level.

The influence depth relating to the average $\overline{N}$ value is calculated from Equation 5.34:

$Z_I = 3.0^{0.75} = 2.3$ m, thus no correction for $\overline{N}$ is needed. Correction for saturated state using Equation 5.35:

$N' = 15 + 0.5(19 - 15) = 17$. Settlement is computed from Equation 5.33:

$S_e(3 \text{ m by } 3 \text{ m}) = 3.0^{0.75} \frac{1.7}{17^{1.4}} \times 200.0 = 14.7$ mm. Shape factor from Equation 5.39:

$I_s = [(1.25 \times 5.0/3.0)/(5.0/3.0 + 0.25)]^2 = 1.18$.
$S_e(3 \text{ m by } 5 \text{ m}) = 14.7 \times 1.18 = 17.3$ mm.

## Problem 5.16

A rectangular footing of width 2 m and length of 20 m is located at the ground surface. It exerts a net uniform contact pressure of 180 kPa to the underlying deep deposit of sand. Calculate the settlement of the footing using Schmertmann's modified strain influence factor diagrams. The average results of the Cone penetration test (*CPT*) are shown in the table below.

| Depth (m) | 1.0 | 2.0 | 3.0 | 4.0 | 5.0 | 6.0 | 7.0 |
|---|---|---|---|---|---|---|---|
| $q_c$ (kPa) | 1600 | 1600 | 2000 | 1800 | 2400 | 2400 | 2600 |

For calculation of $E_s$ use the plane strain conditions expressed by: $E_s = 3.5q_c$.

Solution:

An alternative method of estimating settlement is based on the following equation:

$$S_e = \sum_{i=1}^{i=n} (\varepsilon_z z_l)_i \tag{5.41}$$

where $n$ is the number of the layers within an influence depth of $Z_I$, $(z_l)_i$ is the thickness of a typical layer and $(\varepsilon_z)_i$ is the average vertical strain of the layer. Substituting $(\varepsilon_z)_i$ from Hooke's law (Equations 5.1) into Equation 5.41 we obtain:

$$S_e = \sum_{i=1}^{i=n} \left[ \frac{\sigma_z - \mu(\sigma_x + \sigma_y)}{E_s} z_l \right]_i \tag{5.42}$$

Schmertmann's proposal in sands involves introducing a strain influence factor $I_z$ to replace the stress term in Equation 5.42. The settlement of each layer is defined by:

$$I_z = \frac{\sigma_z - \mu(\sigma_x + \sigma_y)}{q} \rightarrow \Delta S_e = \frac{q \times I_z}{E_s} z_l \tag{5.43}$$

Equation 5.41 may therefore be written as:

$$S_e = q \sum_{0}^{Z_I} \frac{I_z}{E_s} z_l \tag{5.44}$$

The influence depth $Z_I$ is defined by:

$$Z_I = 2B \left( 1 + \log \frac{L}{B} \right) \tag{5.45}$$

The Modulus of Elasticity must be determined for each layer by means of in-situ testing, preferably using the Cone Penetration Test (*CPT*). Strain influence factors, given by Schmertmann (1970) idealize the Boussinesq pattern by straight lines and are based on extensive in-situ tests, and are shown in Figure 5.11. The results obtained from Equation 5.44 must be multiplied by depth and time factors defined by:

$$C_1 = 1 - 0.5(p_o' / q) \geq 0.5 \tag{5.46}$$

$$C_2 = 1 + 0.2 \log(t / 0.1) \tag{5.47}$$

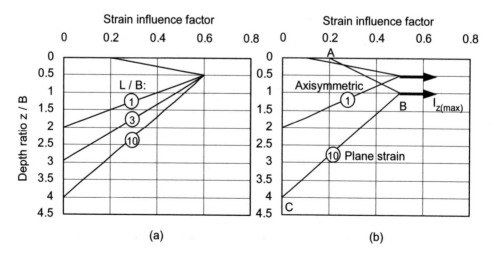

(a)  (b)

Figure 5.11. (a) Strain influence factor (Schmertmann, 1970), (b) improved strain influence factor (Schmertmann, et al., 1978) used in the solution of Problem 5.16.

where $p'_o$ is the effective overburden pressure at the footing level and $t$ is the elapsed time in years. The term $q$ represents the net contact pressure at the foundation level. If the footing is at some depth below the ground surface, $q$ has to be replaced by $q - p'_o$. In the improved strain influence factor diagram the maximum value of $I_z$ is given by:

$$I_{z(max)} = 0.5 + 0.1 \sqrt{\frac{q - p'_o}{p'_{(B/2 \text{ or } B)}}} \tag{5.48}$$

where $p'_{(B/2 \text{ or } B)}$ is the effective overburden pressure at the depth of $B/2$ or $B$ from the footing level depending on the $L/B$ ratio. For $L/B = 1$ (axisymmetric) and $L/B > 10$ (plane strain), the effective overburden pressures corresponds to $B/2$ and $B$ respectively.

Calculate $I_{z(max)}$ from Equation 5.48:

$$I_{z(max)} = 0.5 + 0.1\sqrt{(180.0 - 0.0)/18.0 \times 2.0} = 0.7236.$$

The soil under the footing is divided into 8 layers each having 1 m thickness. The influence factors and the Modulus of Elasticity are then calculated at the centre of each layer. Equations of lines $BA$ and $BC$ (Figure 5.11(b)) are:

$$(BA) \rightarrow \quad \frac{0.7236 - 0.2}{2.0 - 0.0} = \frac{I_z - 0.2}{z - 0.0} \rightarrow I_z = 0.2618z + 0.2.$$

$$(BC) \rightarrow \quad \frac{0.7236 - 0.0}{2.0 - 8.0} = \frac{I_z - 0.0}{z - 8.0} \rightarrow I_z = -0.1206z + 0.9648.$$

The results are summarized in the following table. The $I_z$ values are shown in the third column whilst the forth column shows estimation of $E_s$ values at the mid-point of each layer.

| Layer | $z$ (m) | $I_z$ | $E_s$ (kPa) | $I_z \times z_l / E_s$ (mm/kPa) | $S_e$ (mm) |
|-------|---------|-------|-------------|---------------------------------|------------|
| 1 | 0.5 | 0.3309 | 5600 | $59.1 \times 10^{-3}$ | 10.6 |
| 2 | 1.5 | 0.5927 | 5600 | $105.8 \times 10^{-3}$ | 19.0 |
| 3 | 2.5 | 0.6633 | 6300 | $105.3 \times 10^{-3}$ | 18.9 |
| 4 | 3.5 | 0.5427 | 6650 | $81.6 \times 10^{-3}$ | 14.7 |
| 5 | 4.5 | 0.4221 | 7350 | $57.4 \times 10^{-3}$ | 10.3 |
| 6 | 5.5 | 0.3015 | 8400 | $35.9 \times 10^{-3}$ | 6.5 |
| 7 | 6.5 | 0.1809 | 8750 | $20.7 \times 10^{-3}$ | 3.7 |
| 8 | 7.5 | 0.0603 | 9100 | $6.6 \times 10^{-3}$ | 1.2 |
|   |   |   |   | Total: | $\approx 85$ |

**Problem 5.17**

Calculate the contact pressures under the corners of a rigid square footing using the following data:
$P = 600$ kN, $M_x = -60$ kN.m, $M_y = 100$ kN.m, $L = B = 2$ m.

Solution:

Consider a rigid footing subjected to parallel vertical forces and moments that are applied about the axes in the plane of footing. This system is equivalent to a single resultant force $P$ that is perpendicular to the plane of footing. It is more convenient to express the contact pressure in terms of the coordinate system $Gxyz$ where $G$ represents the centroid of the footing, $z$ is perpendicular to the footing and parallel to the contact pressure, and $xGy$ is the plane of the footing. To simplify the problem the resultant force $P$ is moved from its application point to point $G$, which requires the addition of the two external moments, $M_x$ and $M_y$, to maintain equilibrium:

$$M_x = -P\,y_O, \ M_y = P\,x_O \tag{5.49}$$

where $x_O$ and $y_O$ are the coordinates of point $O$ (assumed both positive) in the $xGy$ system, and the right-hand rule sign convention for moments is adopted. For a footing symmetrical about the $Gx$ or $Gy$ axes the contact pressure is:

$$q = \frac{M_y}{I_y}x - \frac{M_x}{I_x}y + \frac{P}{S} \tag{5.50}$$

where $S$ is the (plan) contact area of the footing, $I_x$ and $I_y$ are the second moments of area of the footing about the $x$ and $y$ axes. For a rigid rectangular footing with the symmetric loading (about the $x$-axis), we can use Equation 5.50 to obtain the contact pressures at the two edges of the rectangle.

$$q_{max} = \frac{P}{LB}\left(1 + \frac{6e}{L}\right) \rightarrow x = L/2, \ q_{min} = \frac{P}{LB}\left(1 - \frac{6e}{L}\right) \rightarrow x = -L/2 \tag{5.51}$$

where $x_O$ is replaced by $e$ to give the familiar expressions used in soil mechanics-geotechnical engineering textbooks.

Using Equation 5.50 with:

$$I_x = I_y = 2.0(2.0)^3/12 = 1.333\text{m}^4,$$

$$q = \frac{100.0}{1.333}(-1.0) - \frac{-60.0}{1.333}(+1.0) + \frac{600.0}{2.0 \times 2.0} = 120.0 \text{ kPa (left-top corner)}.$$

$$q = \frac{100.0}{1.333}(+1.0) - \frac{-60.0}{1.333}(+1.0) + \frac{600.0}{2.0 \times 2.0} = 270.0 \text{ kPa (right-top corner)}.$$

$$q = \frac{100.0}{1.333}(+1.0) - \frac{-60.0}{1.333}(-1.0) + \frac{600.0}{2.0 \times 2.0} = 180.0 \text{ kPa (right-bottom corner)}.$$

$$q = \frac{100.0}{1.333}(-1.0) - \frac{-60.0}{1.333}(-1.0) + \frac{600.0}{2.0 \times 2.0} = 30.0 \text{ kPa (left-bottom corner)}.$$

Problem 5.18

Calculate the contact pressures for the rigid footing of Figure 5.12 at points $a$, $b$ and $c$.

Solution:

Find the centroid: total area of the footing,

$$S = S(abda') + S(b'c'c''d') = S_1 + S_2 = 2.0 \times 1.0 + 2.0 \times 1.0 = 4.0 \text{ m}^2.$$

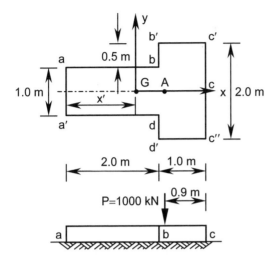

Figure 5.12. Problem 5.18.

Taking the moments of the area about the left edge:

$S_1 \times x_1' + S_2 \times x_2' = S \times x'$, $2.0 \times 1.0 \times 1.0 + 2.0 \times 1.0 \times 2.5 = 4.0 \times x'$,

$x' = 1.75$ m.

Calculate $I_y$:

$$I_y = \frac{1.0 \times 2.0^3}{12} + 1.0 \times 2.0 \times (1.75 - 1.0)^2 + \frac{2.0 \times 1.0^3}{12} + 1.0 \times 2.0 \times (2.5 - 1.75)^2,$$

$I_y = 3.0833 \, \text{m}^4$.

$M_y = 1000.0 \times GA = 1000.0(3.0 - 0.9 - 1.75) = 350.0$ kN.m.

Using Equation 5.50 (note $M_x = 0$):

$$q_a = \frac{350.0}{3.0833}(-1.75) + \frac{1000.0}{4.0} = 51.3 \text{ kPa.}$$

$$q_b = \frac{350.0}{3.0833}(2.0 - 1.75) + \frac{1000.0}{4.0} = 278.4 \text{ kPa.}$$

$$q_c = \frac{350.0}{3.0833}(3.0 - 1.75) + \frac{1000.0}{4.0} = 391.9 \text{ kPa.}$$

Problem 5.19

The following data apply to a square footing located at the ground surface:

$L = B = 1.8$ m, $P = 270$ kN, $M_x = -160$ kN.m, $M_y = 160$ kN.m. Determine:

(a) the contact pressure at the corners of the footing,

(b) the equation of the zero pressure line, its readjusted position (by formulation) and the maximum contact pressure (assume readjustment occurs parallel to the original zero pressure line),

(c) the size of the footing to limit the maximum contact pressure to 280 kPa and the position of the new zero pressure line.

Solution:

(a) Using Equation 5.50 with $I_x = I_y = 1.8(1.8)^3 / 12 = 0.8748 \text{ m}^4$,

$$q = \frac{160.0}{0.8748}(-0.9) - \frac{-160.0}{0.8748}(0.9) + \frac{270.0}{1.8 \times 1.8} = 83.3 \text{ kPa (left-top corner)}.$$

$$q = \frac{160.0}{0.8748}(+0.9) - \frac{-160.0}{0.8748}(+0.9) + \frac{270.0}{1.8 \times 1.8} = 412.5 \text{ kPa (right-top corner)}.$$

$$q = \frac{160.0}{0.8748}(+0.9) - \frac{-160.0}{0.8748}(-0.9) + \frac{270.0}{1.8 \times 1.8} = 83.3 \text{ kPa (right-bottom corner)}.$$

$$q = \frac{160.0}{0.8748}(-0.9) - \frac{-160.0}{0.8748}(-0.9) + \frac{270.0}{1.8 \times 1.8} = -245.9 \text{ kPa (left-bottom corner)}.$$

(b) To find the equation of zero pressure line we set Equation 5.50 to zero:

$$q = \frac{160.0}{0.8748}x - \frac{-160.0}{0.8748}y + \frac{270.0}{1.8 \times 1.8} = 0, \ x + y + 0.4556 = 0.$$

The zero pressure line (*ZPL*) is shown in Figure 5.13 (line *ef*).
To locate the readjusted zero pressure line (*R-ZPL*), we assume it is located at the right side of point *G* (line *kl*) as shown in Figure 5.13. Calculate the second moment of area of triangle *bkl* about its base *kl*:

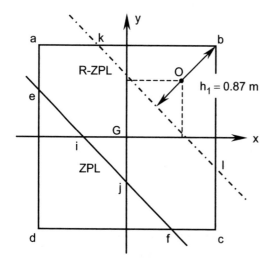

Figure 5.13. Problem 5.19.

$I = \dfrac{bh^3}{12} \rightarrow h = h_1$ and $b = kl = 2h_1$. Thus $I = \dfrac{h_1^4}{6}$.

Calculate the moment of the resultant of external forces about (*R-ZPL*):

The application point of the resultant ($P = 270$ kN) is point $O$ and its distance from (*R-ZPL*) is:

$$h = h_1 - (bG - OG) = h_1 - (L\cos 45.0° - \sqrt{2} \times e_x \text{ or } e_y),$$

$$h = h_1 - (\sqrt{2} \times 1.8/2 - \sqrt{2} \times 160.0/270.0) = h_1 - 0.435 \text{ m}.$$

$$M = 270.0(h_1 - 0.435).$$

$$q_b = \text{contact pressure at } b = \frac{Mh_1}{I} = \frac{270.0(h_1 - 0.435)h_1}{h_1^4/6},$$

$$q_b = \frac{1620.0(h_1 - 0.435)}{h_1^3}.$$

Equate the soil reaction with the applied force (vertical equilibrium):

$$q_b \times \frac{1}{3} \times \text{area of triangle } bkl = \frac{1620.0(h_1 - 0.435)}{h_1^3} \times \frac{1}{3} \times h_1^2 = 270.0 \rightarrow h_1 = 0.87 \text{ m}.$$

$$q_b = \frac{1620.0(0.87 - 0.435)}{0.87^3} \approx 1070 \text{ kPa}.$$

(c) Define parameter $h$ in terms of the length of the square footing $L$:

$$h = h_1 - (\sqrt{2} \times L/2 - \sqrt{2} \times 160.0/270.0) = h_1 - (0.707L - 0.838) \text{ m}.$$

$$M = 270.0\big[h_1 - (0.707L - 0.838)\big], \quad q_b = \frac{Mh_1}{I} = \frac{270.0 \times 6\big[h_1 - (0.707L - 0.838)\big] \times h_1}{h_1^4},$$

$$q_b = \frac{270.0 \times 6\big[h_1 - (0.707L - 0.838)\big]}{h_1^3} = 280.0. \text{ Vertical equilibrium:}$$

$$q_b \times \frac{1}{3} \times \text{area of triangle } bkl = 280.0 \times \frac{1}{3} \times h_1^2 = 270.0 \rightarrow h_1 = 1.7 \text{ m}. \text{ Substituting } h_1 \text{ in } q_b:$$

$$q_b = \frac{270.0 \times 6\big[1.7 - (0.707L - 0.838)\big]}{1.7^3} = 280.0 \rightarrow L = 2.4 \text{ m}.$$

$h_1 \approx (\sqrt{2}/2)L$. Thus the new zero pressure line is $ac$ (Figure 5.13).

**Problem 5.20**

For a trapezoidal rigid footing shown in Figure 5.14 compute the dimensions $B$ and $A$ that will ensure a uniform contact pressure of 200 kPa.

Solution:

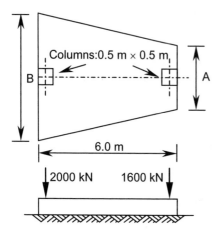

Figure 5.14. Problem 5.20.

Locate the application point of the soil reaction. Taking moments about the left edge of the footing:

$$2000.0 \times 0.25 + 1600.0 \times 5.75 = (2000.0 + 1600.0)x',$$

$$x' = \frac{9700.0}{3600.0} = 2.694 \text{ m}.$$

For a uniform contact pressure this point must be the centroid of the contact area:

$$x' = \frac{L}{3}\frac{2A+B}{A+B} = \frac{6.0}{3}\frac{2A+B}{A+B} = 2.694.$$

$$\frac{B}{A} = 1.882.$$

For a uniform contact pressure of 200 kPa:

$$\frac{A+B}{2}6.0 \times 200.0 = 3600.0 \rightarrow A+B = \frac{3600.0 \times 2}{6.0 \times 200.0} = 6.0.$$

Solving for $A$ and $B$:
$A = 2.08$ m, $B = 3.92$ m.

Problem 5.21

Predict the soil settlement and reaction at selected points for the finite beam shown in Figure 5.15. The thickness of the footing is 0.7 m and its width is 1 m. The beam is divided into 5 equal segments of 2 m each.
$E_f = 22 \times 10^6$ kPa, $E_s = 10440$ kPa, $\mu = 0.3$.

Solution:

The modulus of subgrade reaction is defined as:

$$k_s = q/S_e \tag{5.52}$$

where $q$ is the contact pressure under a loading plate and $S_e$ is the corresponding settlement. This modulus is measured in kN/m³.

The relationship between $k_s$ and Modulus of Elasticity $E_s$ has been extensively investigated, and the Vesić equation is often used in design:

$$k'_s = 0.65 \times 12\sqrt{\frac{E_s B^4}{E_f I_f}} \frac{E_s}{1-\mu^2}$$  (5.53)

where $E_s$, $E_f$ are the Moduli of Elasticity of soil and footing respectively, $B$ is the width of the footing, $I_f$ is the second moment of area of the footing section, $\mu$ is Poisson's ratio for the soil, and $k'_s = k_s B$.

A simplified version of Equation 5.53 is also used:

$$k_s = \frac{E_s}{B(1-\mu^2)}$$  (5.54)

The Winkler model (beam resting on bed of springs) can be simplified by replacing the bed of springs with a finite number of springs of stiffness $K$ where:

$$K = \text{contact area corresponding to a single spring} \times k_s = B \times a \times k_s = k'_s a$$  (5.55)

where $a$ is the horizontal distance between the parallel springs; thus

$$R_i = K z_i = k'_s a z_i = k_s B a z_i$$  (5.56)

The governing differential equation for the deflection of the beam is:

$$\frac{d^2 z}{dx^2} = \frac{-M(x)}{EI}$$  (5.57)

where $M(x)$ is the bending moment at distance $x$ from the origin (say left edge) and $EI$ is the flexural stiffness of the beam. Using Equation 5.53 to calculate $k'_s$:

$$k'_s = 0.65 \times 12\sqrt{\frac{10440 \times 1.0^4}{22 \times 10^6 (0.7)^3 \times 1.0/12}} \times \frac{10440}{1 - 0.3^2} = 5300 \text{ kPa},$$

$K = k'_s \times a = 5300 \times 2.0 = 10600$ kN/m. For the edges: $K = k'_s \times (a/2) = 5300$ kN/m.

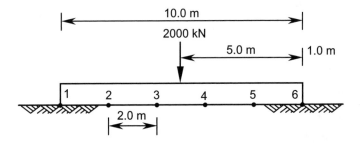

Figure 5.15. Problem 5.21.

Thus, $R_1 = 5300\ z_1$, $R_2 = 10600\ z_2$, $R_3 = 10600\ z_3$. Because of the symmetrical loading, the required parameters at points 1, 2 and 3 are equal to the corresponding values at points 6, 5, and 4. To solve the problem three linear equations involving the three unknowns of $z_1$, $z_2$ and $z_3$ (settlements at points 1, 2 and 3) must be developed. Applying the finite difference equations at points 2 and 3, we have:

$$(\frac{d^2 z}{dx^2})_i = \frac{z_{i+1} + z_{i-1} - 2z_i}{a^2} = \frac{-M_i}{EI},$$

$$(\frac{d^2 z}{dx^2})_2 = \frac{z_3 + z_1 - 2z_2}{2.0^2} = \frac{-M_2}{EI},$$

$$(\frac{d^2 z}{dx^2})_3 = \frac{z_4 + z_2 - 2z_3}{2.0^2} = \frac{z_2 - z_3}{2.0^2} = \frac{-M_3}{EI}.$$

Note $z_3 = z_4$.

$EI = 22.0 \times 10^6 \times (1.0 \times 0.7^3 / 12) = 628833.33$ kN.m$^2$.

Calculate bending moments at points 2 and 3:

$M_2 = R_1 \times 2.0 = 10600z_1$,

$M_3 = R_1 \times 4.0 + R_2 \times 2.0 = 21200z_1 + 21200z_2$.

Substituting $M_2$ and $M_3$ into the finite difference equations:

$$\frac{z_3 + z_1 - 2z_2}{4.0} = \frac{-10600z_1}{628833.33},$$

$$\frac{z_2 - z_3}{4.0} = \frac{-(21200z_1 + 21200z_2)}{628833.33}.$$

The third equation is obtained from (static) equilibrium of the vertical forces:

$2R_1 + 2R_2 + 2R_3 = 2000.0$,

$R_1 + R_2 + R_3 = 1000.0$ or:

$5300\ z_1 + 10600z_2 + 10600z_3 = 1000.0$.

Simplifying and rearranging the equations:

$671233.33z_1 - 1257666.66z_2 + 628833.33z_3 = 0.0$,

$84800z_1 + 713633.33z_2 - 628833.33z_3 = 0.0$,

$5300z_1 + 10600z_2 + 10600z_3 = 1000.0$.

Solving for $z_1$, $z_2$, and $z_3$:

$z_1 = 26.2 \times 10^{-3}$ m = 26.2 mm.

$z_2 = 36.4 \times 10^{-3}$ m = 36.4 mm.

$z_3 = 44.8 \times 10^{-3}$ m = 44.8 mm.

$R_1 = 5300 \times 26.2 \times 10^{-3} \approx 139$ kN. ↑

$R_2 = 10600 \times 36.4 \times 10^{-3} \approx 386$ kN. ↑

$R_3 = 10600 \times 44.8 \times 10^{-3} \approx 475$ kN. ↑

## 5.3 REFERENCES AND RECOMMENDED READINGS

ACI, 1988. Suggested analysis and design procedures for combined footings and mats. *AC structural journal*, 85(3): 304-324.

Ahlvin, R.G. & Ulery, H.H.1962. Tabulated values for determining the complete pattern of stresses, strains, and deflections beneath a uniform load on a homogeneous half space. *Highway research board bulletin*, (342): 1-13.

Aysen, A. 2002. *Soil mechanics: Basic concepts and engineering applications*. Lisse: Balkema.

Bowles, J.E. 1996. *Foundation analysis and design*. 5th edition. New York: McGraw-Hill.

Burland, J.B. 1970. Discussion. *Proc. conf. on in-situ investigations in soils and rocks*. London: British Geotechnical Society.

Burland, J.B. & Burbidge, M.C. 1985. Settlements of foundations on sands and gravel. *Proc.*, Part 1, 78: 1325-1381. London: Institution of Civil Engineers.

Christian, J.T. & Carrier, W.D. 1978. Janbu, Bjerrum, and Kjaernsli's chart reinterpreted. *Canadian geotechnical journal*, 15(1): 123, 436.

Duncan, J.M. & Buchignani, A.L. 1976. *An engineering manual for settlement studies*. Berkeley: University of California.

Fadum, R.E. 1948. Influence values for estimating stresses in elastic foundations. *Proc. 2nd intern. conf. SMFE*, 3: 77-84. Rotterdam.

Fox, E.N. 1948. The mean elastic settlement of a uniformly loaded area at a depth below the ground surface. *Proc. 2nd intern. conf. SMFE*, 1: 129-132. Rotterdam.

Geddes, J.D. 1966. Stresses in foundation soils due to vertical subsurface loading. *Geotechnique*, 16(3): 231-255.

Geddes, J.D. 1969. Boussinesq based approximations to the vertical stresses caused by pile type subsurface loadings. *Geotechnique*, 19(4): 509-514.

Giroud, J.P. 1972. Settlement of rectangular foundation on soil layer. *Journal SMFED, ASCE*, 98(SM1): 149-154.

Harr, M.E. 1966. *Foundations of theoretical soil mechanics*. New York: McGraw-Hill.

Holtz, A.D.1991. Stress distribution and settlement of shallow foundations. In H.Y. Fang (ed.), *Foundation engineering handbook*. New York: Van Nostrad Reinhold.

Janbu, N., Bjerrum, L., & Kjaernsli, B. 1956. Veileduing ved losning, av fundamenteringsopogaver (Soil mechanics applied to some engineering problems). *Noewegian geotechnical institiute* Pub. 16. Oslo.

Jumikis, A.R. 1969. *Theoretical soil mechanics*. New York: Van Nostrand Reinhold.

Liao, S.S.C. 1995. Estimating the coefficient of subgrade reaction for plane strain conditions. *Proc. institution of civil engineering, geotecnical engineering*, 113: 166-181.

Mayne, P.W. & Poulos, H.G. 1999. Approximate displacement influence factors for elastic shallow foundations. *Journal GE, ASCE*, 129(6): 453-460.

Meyerhof, G.G. & Fellenius, B.H. 1985. *Canadian foundation engineering manual*. 2nd edition. Canada: Canadian Geotechnical Society.

Mindlin, R.D. 1936. Force at a point in the interior of a semi-infinite solid. *Journal of the American institution of physics*, 7(5): 195-202.

Newmark, N.M. 1942. Influence charts for computations of stresses in elastic soils. *Engineering experimental station bulletin* (367). University of Illinois.

Peck, R.B., Hanson, W.E. & Thornburn, T.H. 1974. *Foundation engineering*. 2nd edition. New York: John Wiley & Sons.

Poulos, H.G. & Davis, E.A. 1974. *Elastic solutions for soil and rock mechanics*. New York: Wiley.

Powrie, W. 1997. *Soil mechanics-concepts and applications*. London: E & FN Spon.

Schmertmann, J.H. 1970. Static cone to compute static settlement over sand. *Journal SMFED, ASCE*, 96(SM3): 1011-1043.

Schmertmann, J.H., Hartman, J.P. & Brown, P.R. 1978. Improved strain influence factor diagrams. *Journal GE, ASCE*, 104(8): 1131-1135.

Steinbrenner, W. 1934. Tafeln zur Setzungsberechnung. *Die strasse* 1.

Terzaghi, K., Peck, R. B., & Mesri, G. 1996. *Soil mechanics in engineering practice*. 3rd edition. New York: John Wiley & Sons.

Tomilinson, M.J. 1995. *Foundation design and construction*. 6th edition. Harlow, Essex: Longman Scientific & Technical.

Ulrich, E.J., Jr. 1991. Subgrade reaction in mat foundation design. *Concrete international*, 13(1): 41-50.

Ulrich, E.J., Jr. 1994. Mat foundation design: A historical perspective. *Proc. ASCE speciality conf. on vertical and horizontal deformations of foundations and embankments*, 107-120. Texas A & M, College Station 1.

Vesić, A.S. 1961a. Beams on elastic subgrade and the Winkler's hypothesis. *Proc. 5th intern. conf. SMFE*, 1: 845-850.

Vesić, A.S. 1961b. Bending of beams resting on isotropic elastic solid. *Journal EMD, ASCE*, 87(2): 35-53.

CHAPTER 6

# One Dimensional Consolidation

## 6.1  INTRODUCTION

The problems solved in this chapter are related to the prediction of both magnitude and rate of one-dimensional consolidation settlement including the determination of consolidation characteristics in the laboratory and in the field.

The parameters calculated from the laboratory results include the *compression index* $C_c$, the *coefficient of volume compressibility* $m_v$, the *preconsolidation pressure* $p'_c$, and the *coefficient of consolidation* $c_v$. Parameters $C_c$ and $m_v$ represent the basic load-deformation models. These models are based on the laboratory results that are shown in $e$-$\log\sigma'$ ($e$ being the void ratio) coordinate system for $C_c$, and $\Delta H/H$ ($\Delta H$ being the vertical deformation of the laboratory sample)-$\sigma'$ coordinate system for $m_v$. For the effective vertical stresses greater than preconsolidation pressure the $e$-$\log\sigma'$ curve obtained in the laboratory test can be approximated by a straight line (*virgin compression line*) where parameter $C_c$ defines the slope of this line. The parameter $m_v$ is the slope of the $\Delta H/H$-$\sigma'$ curve and, which changes with the stress level.

The determination of the above two parameters are described in Problems 6.1 and 6.4. Estimation of $C_c$ in the field and other consolidation characteristics are explained in Problem 6.9.

Assuming that the field values of $C_c$ and $m_v$ are identical to the values obtained in the laboratory, both parameters may be used to calculate the vertical deformation of a finite layer within a given time span (Problems 6.2, 6.5, and 6.6).

In order to estimate the progress of consolidation settlement with time (rate of consolidation) it is necessary to evaluate the coefficient of consolidation $c_v$ from the laboratory consolidation test (Problem 6.3) or from the field based measurement data (Problem 6.9). The mathematical relationship between the rate of consolidation defined by the average degree of consolidation and time can also be approximated by parabolic equations; this is explained in Problem 6.7.

Lowering the water table in the field (by pumping) induces an increase in the vertical effective stresses and causes the consolidation settlement as described in Problem 6.8.

Problem 6.10 explains a simplified design for vertical drains. The vertical drains facilitate the horizontal flow of water and thereby accelerate consolidation process.

In the derivation of the mathematical relationships for one-dimensional consolidation it is common to assume that the increase in the effective vertical stress is instantaneous. However, construction work is normally a gradual process and may be idealized as linear function of time as explained in Problem 6.11.

## 6.2 PROBLEMS

Problem 6.1

Data obtained from a laboratory consolidation test are tabulated as follows:

| Test points | 1 | 2 | 3 | 4 | 5 | 6 |
|---|---|---|---|---|---|---|
| $\sigma'$ (kPa) | 20 | 50 | 100 | 200 | 400 | 800 |
| Total $\Delta H$ (mm) | 0.23 | 0.87 | 1.90 | 3.62 | 5.55 | 7.25 |

$G_s = 2.70$, $H_0 =$ (initial thickness at zero pressure) $= 22.5$ mm,
$w =$ (moisture content at the beginning of the test) $= 0.78$.
Plot the e-log$\sigma'$ curve and calculate $C_c$.

Solution:

The compression index $C_c$ is the absolute value of the gradient of the virgin compression line:

$$C_c = \frac{e_0 - e_1}{\log\sigma'_1 - \log\sigma'_0} = \frac{e_0 - e_1}{\log(\sigma'_1/\sigma'_0)} \tag{6.1}$$

where the subscripts 0 and 1 represent the initial and final states of void ratios and corresponding effective vertical stresses. In a fully saturated soil undergoing a vertical deformation $\Delta H$, the following can be derived from the phase diagram (Chapter 1):

$$\frac{\Delta e}{\Delta H} = \frac{1+e_0}{H_0} \tag{6.2}$$

In saturated soil:
$e = wG_s$; thus $e_0 = 0.78 \times 2.7 = 2.106$ (at zero pressure).
Using Equation 6.2 the corresponding void ratios are calculated and shown in Figure 6.1.

$$\frac{2.106 - e_1}{0.23} = \frac{1+2.106}{22.5} \rightarrow e_1 = 2.074.$$

$$\frac{2.106 - e_2}{0.87} = \frac{1+2.106}{22.5} \rightarrow e_2 = 1.986.$$

$$\frac{2.106 - e_3}{1.90} = \frac{1+2.106}{22.5} \rightarrow e_3 = 1.844.$$

$$\frac{2.106 - e_4}{3.62} = \frac{1+2.106}{22.5} \rightarrow e_4 = 1.606.$$

$$\frac{2.106 - e_5}{5.55} = \frac{1+2.106}{22.5} \rightarrow e_5 = 1.340.$$

$$\frac{2.106 - e_6}{7.25} = \frac{1+2.106}{22.5} \rightarrow e_6 = 1.105.$$

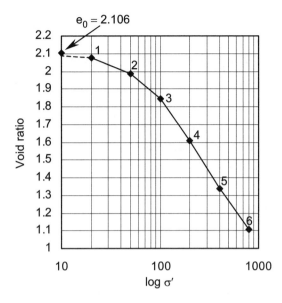

Figure 6.1. Problem 6.1.

The value of $C_c$ is estimated by assuming a line through test points 4 and 6, therefore using Equation 6.1:

$$C_c = \frac{1.606 - 1.105}{\log 800.0 - \log 200.0} = 0.832.$$

Problem 6.2

An open layer of clay 4 m thick is subjected to loading that increases the average effective vertical stress from 185 kPa to 310 kPa. Determine:
(a) the total settlement,
(b) the settlement at the end of 1 year,
(c) the time in days for 50% consolidation,
(d) the time in days for 25 mm of settlement to occur.
$m_v = 0.00025$ m$^2$/kN, $c_v = 0.75$ m$^2$/year.

Solution:

(a) Combining Equations 6.1 and 6.2 and replacing $\Delta H$ by consolidation settlement (in the field) $S_c$ the following equation is obtained:

$$S_c = \frac{C_c H}{1 + e_0} \log \frac{\sigma_1'}{\sigma_0'} \tag{6.3}$$

where $H$ is the initial thickness of the clay layer and $\sigma_0'$ and $\sigma_1'$ represent the average effective vertical stresses within the layer at initial and final states respectively. The compressibility of a clay layer can also be represented by the coefficient of volume compressibility $m_v$ defined by:

$$m_v = \frac{\Delta H / H_0}{\sigma_1' - \sigma_0'} \quad \text{or} \quad m_v = \frac{\Delta e /(1+e_0)}{\sigma_1' - \sigma_0'} \tag{6.4}$$

From this definition we can find the consolidation settlement for a layer of thickness $H$:

$$S_c = \Delta H = m_v H (\sigma_1' - \sigma_0') \tag{6.5}$$

Using the above equation:

$S_c = 0.00025 \times 4.0 (310.0 - 185.0) = 0.125$ m $= 125$ mm.

(b) A dimensionless variable called the time factor $T_v$ is defined as:

$$T_v = \frac{c_v t}{d^2} \tag{6.6}$$

The parameter $c_v$ is called the coefficient of consolidation and is derived by considering the flow continuity:

$$c_v = \frac{k}{m_v \gamma_w} \tag{6.7}$$

where $k$ is the coefficient of permeability and $\gamma_w$ is the unit weight of water. The parameter $d$ represents the maximum drainage length that a particle of water has to cover to reach to the free draining boundary. The degree of consolidation at any depth is:

$$U_z = \frac{u_i - u_e}{u_i} = 1 - \frac{u_e}{u_i} \tag{6.8}$$

where $u_i$ and $u_e$ are the initial and excess pore pressures (due to loading) at depth $z$ respectively. The average value of degree of consolidation for a layer can be expressed in terms of the time factor $T_v$ by:

$$U = 1 - \sum_{m=0}^{m=\infty} \frac{2}{M^2} \exp\left(- M^2 T_v\right) \tag{6.9}$$

where $M = \pi(2m+1)/2$. $U$ may be approximated by the following:

$$T_v = \pi U^2 / 4 \rightarrow U \le 0.6, \quad Tv = -0.933 \log(1-U) - 0.085 \rightarrow U \ge 0.6 \tag{6.10}$$

It can be shown that the average degree of consolidation is:

$$U = \frac{S}{S_c} \tag{6.11}$$

where $S$ is the settlement at time $t$ and $S_c$ is the final settlement ($t \rightarrow \infty$).

The step procedure for calculation of the settlement at a specified time includes:
1. Calculate time factor $T_v$ from Equation 6.6.
2. Calculate average degree of consolidation $U$ from Equations 6.9 or 6.10.
3. Use Equation 6.11 for calculation of $S$. Thus

1. $T_v = \dfrac{0.75 \times 1.0}{(4.0/2)^2} = 0.1875.$

2. $T_v = \pi U^2 / 4 = 0.1875 \rightarrow U = 0.4886 < 0.6.$

3. $U = S/125.0 = 0.4886 \rightarrow S = 61.0$ mm.

(c) For 50% consolidation $T_v = 0.197$, therefore:

$$0.197 = \frac{0.75 \times t}{(4.0/2)^2} \rightarrow t = 1.0507 \text{ year} = 383.5 \text{ days.}$$

(d) $U = 25.0/125.0 = 0.2, T_v = \pi U^2/4 = \pi \times 0.2^2/4 = 0.0314$,

$$T_v = \frac{0.75 \times t}{(4.0/2)^2} = 0.0314 \rightarrow t = 0.1675 \text{ year} = 61.0 \text{ days.}$$

Problem 6.3

Data obtained from a laboratory consolidation test are shown in the table below:

| Time (min) | 0.25 | 1 | 4 | 9 | 16 | 25 | 36 | 81 | 1440 |
|---|---|---|---|---|---|---|---|---|---|
| Total $\Delta H$ (mm) | 0.622 | 1.244 | 2.468 | 3.400 | 3.838 | 3.970 | 4.000 | 4.051 | 4.100 |

$\sigma'_0 = 100$ kPa, $\sigma'_1 = 200$ kPa, $H_0 = 23.6$ mm. Determine:

(a) $c_v$ from the root time plot in $m^2$/year,

(b) $c_v$ from the log time plot in $m^2$/year,

(c) $k$ in m/s.

Solution:

(a) $H_{av}$ (average thickness) $= 23.6 - 4.1/2 = 21.55$ mm, $d = 21.55/2 = 10.775$ mm.

The square root time plot is shown in Figure 6.2(a) from which

$\sqrt{t_{90}} = 3.25$, $t_{90} = 10.56$ min. At 90% consolidation $T_v = 0.848$; thus from Equation 6.6:

$$c_v = 0.848 \times 10.775^2 \times 10^{-6}/[10.56/(365 \times 1440)] = 4.9 \text{ m}^2/\text{year.}$$

(b) The log time plot is shown in Figure 6.2(b) from which $t_{50} = 2.35$ min.

$$c_v = 0.197 \times 10.775^2 \times 10^{-6}/[2.35/(365 \times 1440)] = 5.1 \text{ m}^2/\text{year.}$$

(c) Calculate $m_v$:

From Equation 6.4: $m_v = \dfrac{\Delta H/H_0}{\sigma'_1 - \sigma'_0} = \dfrac{4.1/23.6}{200.0 - 100.0} = 1.737 \times 10^{-3} \text{ m}^2/\text{kN.}$

Using Equation 6.7 with an average value of $c_v = 5$ $m^2$/year:

$$k = c_v m_v \gamma_w = 5.0 \times 1.737 \times 10^{-3} \times 9.81/(365 \times 24 \times 60 \times 60) = 2.7 \times 10^{-9} \text{ m/s.}$$

Problem 6.4

In a one-dimensional consolidation test the time required for 50% consolidation has been measured at 154 seconds (through the observation and measurement of pore water pressure). The settlement of the sample at the end of the test was 2.5 mm.

$\sigma'_0 = 60$ kPa, $\sigma'_1 = 120$ kPa, $e_0 = 0.65$, $H_0 = 20$ mm. Determine:

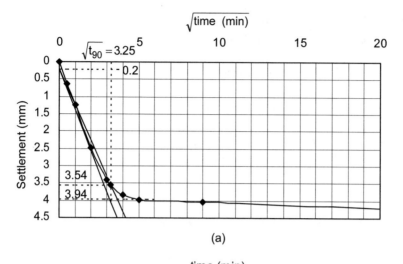

(a)

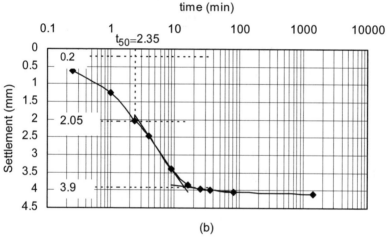

(b)

Figure 6.2. Problem 6.3: (a) square root time method, (b) log time method.

(a) the time required for 90% consolidation,
(b) the coefficient of permeability in m/s,
(c) the compression index.

Solution:

(a) Using Equation 6.6 for 50% and 90% consolidation:

$$0.197 = \frac{c_v \times 154.0}{d^2},$$

$$0.848 = \frac{c_v \times t_{90}}{d^2}.$$

Combining the above two equations:

$$\frac{0.197}{0.848} = \frac{c_v \times 154.0}{d^2} \times \frac{d^2}{c_v \times t_{90}},$$

$$\frac{0.197}{0.848} = \frac{154.0}{t_{90}} \rightarrow t_{90} \approx 663 \text{ s.}$$

(b) Calculate $c_v$ and $m_v$:

$H_{av} = 20.0 - 2.5/2 = 18.75 \text{ mm}, d = 18.75/2 = 9.375 \text{ mm.}$

From Equation 6.6: $c_v = \dfrac{0.197 \times 9.375^2 \times 10^{-6}}{154.0} = 1.124 \times 10^{-7} \text{ m}^2/\text{s.}$

From Equation 6.4: $m_v = \dfrac{\Delta H / H_0}{\sigma_1' - \sigma_0'} = \dfrac{2.5/20.0}{120.0 - 60.0} = 2.083 \times 10^{-3} \text{ m}^2/\text{kN.}$

Using Equation 6.7:

$k = c_v m_v \gamma_w = 1.124 \times 10^{-7} \times 2.083 \times 10^{-3} \times 9.81 = 2.3 \times 10^{-9} \text{ m/s.}$

(c) From Equation 6.2:

$$\frac{\Delta e}{2.5} = \frac{1 + 0.65}{20.0} \rightarrow \Delta e = 0.2062.$$

Thus the compression index is found from Equation 6.1:

$$C_c = \frac{0.2062}{\log(120.0 / 60.0)} = 0.685.$$

Problem 6.5

For a 4 m layer of the clay of Problem 6.4, how long would it take to reach 50% degree of consolidation under the same drainage, physical and stress conditions? What will be the settlement of the clay layer at this stage?

Solution:

Using Equation 6.6: $c_v = 1.124 \times 10^{-7} \times 24 \times 60 \times 60 = \dfrac{0.197 \times 2.0^2}{t} \rightarrow t = 81.1$ days.

The final consolidation settlement is calculated from Equation 6.3:

$$S_c = \frac{0.685 \times 4.0}{1 + 0.65} \log \frac{120.0}{60.0} = 0.5 \text{ m} = 500 \text{ mm.}$$

Therefore: $S(50\%) = 500.0 \times 0.5 = 250.0$ mm.

Problem 6.6

A surface load of 60 kPa is applied on the ground surface over a large area. The soil profile consists of a sand layer 2 m thick, the top of which is the ground surface, overlying a 4 m thick layer of clay. An impermeable boundary is located at the base of the clay layer. The water table is 1 m below the ground surface. If the preconsolidation pressure for a sample of soil from the mid-point of the clay layer is 60 kPa, calculate the consolidation settlement of the clay layer. The properties of the soil section are: sand: $\rho_{dry} = 1.6 \text{ Mg/m}^3$, $\rho_{sat} = 1.9 \text{ Mg/m}^3$, clay: $\rho_{sat} = 1.65 \text{ Mg/m}^3$, $e_0 = 1.5$, $C_c = 0.6$, $C_r = 0.1$.

Solution:

If the initial effective vertical stress $\sigma'_0$ is less than the preconsolidation pressure (overconsolidated state) and the final state is on the virgin compression line (normally consolidated state), then the settlement is calculated in two stages according:

$$S_c = \frac{C_r H}{1+e_0}\log\frac{p'_c}{\sigma'_0}+\frac{C_c H}{1+e_p}\log\frac{\sigma'_1}{p'_c} \qquad (6.12)$$

where $e_p$ is the void ratio corresponding to the preconsolidation pressure and $C_r$ is the recompression index, which can be obtained from either the overconsolidated segment of the $e$-log$\sigma'$ plot or an unloading-reloading cycle.

$\sigma'_0 = 1.65\times2.0\times9.81+1.9\times1.0\times9.81+1.6\times1.0\times9.81-1.0\times3.0\times9.81 = 37.3$ kPa.

$\sigma'_1 = 37.3 + 60.0 = 97.3$ kPa.

In the $e$-log$\sigma'$ coordinate system the equation of the recompression line is:

$e = 1.5-0.1(\log\sigma'-\log37.3)$, thus

$e_p = 1.5-0.1(\log60.0-\log37.3) = 1.479$. Using Equation 6.12:

$$S_c = \frac{0.1\times4.0}{1+1.5}\log\frac{60.0}{37.3}+\frac{0.6\times4.0}{1+1.479}\log\frac{97.3}{60.0},$$

$S_c = 0.033 + 0.203 = 0.236$ m $= 236$ mm.


Problem 6.7

A soil profile consists of a sand layer 2 m thick, whose top is the ground surface, and a clay layer 3 m thick with an impermeable boundary located at its base. The water table is at the ground surface. A widespread load of 100kPa is applied at the ground surface. Construct isochrones corresponding to 10%, 50%, and 90% consolidation, indicate the amount of excess pore pressures on the impermeable boundary and determine the amount of settlement after 2 years. Assume that the soil is in a normally consolidate state. The properties of the soil section are:

sand: $\gamma_{sat} = 20$ kN/m³, clay: $\gamma_{sat} = 16$ kN/m³, $e_0 = 1.3$, $C_c = 0.5$, $c_v = 6.5$ m²/year.

Solution:

The distribution of excess pore pressure in depth is according:

$$u_e = \sum_{m=0}^{m=\infty}\frac{2u_i}{M}\left(\sin\frac{Mz}{d}\right)\exp\left(-M^2 T_v\right) \qquad (6.13)$$

The definitions of $u_i$ and $M$ are described in the explanations of Equations 6.8 and 6.9. The magnitudes of $T_v$ for 10%, 50% and 90% consolidation are calculated from Equations 6.10 and are 0.008, 0.197 and 0.848 respectively. Using Equation 6.13 the following 3 tables are constructed for each $T_v$. The magnitudes of excess pore pressures on the impermeable boundary are found to be 100.0 kPa, 77.8 kPa and 15.7 kPa for $T_v$ values of 0.008, 0.197 and 0.848 respectively. The corresponding isochrones are shown in Figure 6.3. To predict the settlement after 2 years we first find the final settlement:

$\sigma'_0$ (at the centre of clay layer)$= 16.0\times1.5+20.0\times2.0-(1.5+2.0)\times9.81 = 29.7$ kPa.

| $m$ | $M$ | $2/M$ | Exp $(-M^2 T_v)$ | | |
|---|---|---|---|---|---|
| | | | $T_v = 0.008$ | $T_v = 0.197$ | $T_v = 0.848$ |
| 0 | $0.5\,\pi$ | 1.2732 | 0.9804 | 0.6150 | 0.1234 |
| 1 | $1.5\,\pi$ | 0.4244 | 0.8372 | 0.0126 | 0.0000 |
| 2 | $2.5\,\pi$ | 0.2546 | 0.6105 | 0.0000 | 0.0000 |
| 3 | $3.5\,\pi$ | 0.1819 | 0.3801 | 0.0000 | 0.0000 |
| 4 | $4.5\,\pi$ | 0.1415 | 0.2021 | 0.0000 | 0.0000 |
| 5 | $5.5\,\pi$ | 0.1157 | 0.0918 | 0.0000 | 0.0000 |
| 6 | $6.5\,\pi$ | 0.0979 | 0.0356 | 0.0000 | 0.0000 |
| 7 | $7.5\,\pi$ | 0.0849 | 0.0118 | 0.0000 | 0.0000 |

| $m$ | $\sin(Mz/d)$ | | | |
|---|---|---|---|---|
| $\downarrow$ | $z/d$ | | | |
| | 0.250 | 0.500 | 0.750 | 1.000 |
| 0 | 0.3827 | 0.7071 | 0.9239 | 1.0000 |
| 1 | 0.9239 | 0.7071 | -0.3827 | -1.0000 |
| 2 | 0.9239 | -0.7071 | -0.3827 | 1.0000 |
| 3 | 0.3827 | -0.7071 | 0.9239 | -1.0000 |
| 4 | -0.3827 | 0.7071 | -0.9239 | 1.0000 |
| 5 | -0.9239 | 0.7071 | 0.3827 | -1.0000 |
| 6 | -0.9239 | -0.7071 | 0.3827 | 1.0000 |
| 7 | -0.3827 | -0.7071 | -0.9239 | -1.0000 |

| $z/d$ | 0.250 | 0.500 | 0.750 | 1.000 |
|---|---|---|---|---|
| | $z = 0.75$ m | $z = 1.5$ m | $z = 2.25$ m | $z = 3.0$ m |
| $T_v = 0.008$ | | | | |
| $u_{(0)}/u_i$ | $477.703\times10^{-3}$ | $882.634\times10^{-3}$ | $1153.254\times10^{-3}$ | $1248.245\times10^{-3}$ |
| $u_{(1)}/u_i$ | $328.268\times10^{-3}$ | $251.238\times10^{-3}$ | $-135.976\times10^{-3}$ | $-355.307\times10^{-3}$ |
| $u_{(2)}/u_i$ | $143.605\times10^{-3}$ | $-109.907\times10^{-3}$ | $-59.484\times10^{-3}$ | $155.433\times10^{-3}$ |
| $u_{(3)}/u_i$ | $26.460\times10^{-3}$ | $-48.889\times10^{-3}$ | $63.879\times10^{-3}$ | $-69.140\times10^{-3}$ |
| $u_{(4)}/u_i$ | $-10.944\times10^{-3}$ | $20.221\times10^{-3}$ | $-26.421\times10^{-3}$ | $28.597\times10^{-3}$ |
| $u_{(5)}/u_I$ | $-9.813\times10^{-3}$ | $7.510\times10^{-3}$ | $4.065\times10^{-3}$ | $-10.621\times10^{-3}$ |
| $u_{(6)}/u_I$ | $-3.220\times10^{-3}$ | $-2.464\times10^{-3}$ | $1.334\times10^{-3}$ | $3.485\times10^{-3}$ |
| $u_{(7)}/u_I$ | $-0.383\times10^{-3}$ | $-0.708\times10^{-3}$ | $-0.926\times10^{-3}$ | $-1.002\times10^{-3}$ |
| $u_e/u_i$ | $951.676\times10^{-3}$ | $999.635\times10^{-3}$ | $999.725\times10^{-3}$ | $999.690\times10^{-3}$ |
| $U_z$ | 0.0483 | 0.0004 | 0.0003 | 0.0003 |
| $u_e$ (kPa) | 95.17 | 99.96 | 99.97 | 99.97 |

| $z/d$ | 0.250 $z = 0.75$ m | 0.500 $z = 1.5$ m | 0.750 $z = 2.25$ m | 1.000 $z = 3.0$ m |
|---|---|---|---|---|
| $T_v = 0.197$ | | | | |
| $u_{(0)}/u_i$ | $299.661\times10^{-3}$ | $553.672\times10^{-3}$ | $723.430\times10^{-3}$ | $783.018\times10^{-3}$ |
| $u_{(1)}/u_i$ | $4.940\times10^{-3}$ | $3.781\times10^{-3}$ | $-2.046\times10^{-3}$ | $-5.347\times10^{-3}$ |
| $u_e/u_i$ | $304.601\times10^{-3}$ | $557.453\times10^{-3}$ | $721.384\times10^{-3}$ | $777.671\times10^{-3}$ |
| $U_z$ | 0.6954 | 0.4425 | 0.2786 | 0.2223 |
| $u_e$ (kPa) | 30.46 | 55.74 | 72.14 | 77.77 |

| $z/d$ | 0.250 $z = 0.75$ m | 0.500 $z = 1.5$ m | 0.750 $z = 2.25$ m | 1.000 $z = 3.0$ m |
|---|---|---|---|---|
| $T_v = 0.848$ | | | | |
| $u_{(0)}/u_i$ | $60.127\times10^{-3}$ | $111.094\times10^{-3}$ | $145.157\times10^{-3}$ | $157.113\times10^{-3}$ |
| $u_e/u_i$ | $60.127\times10^{-3}$ | $111.094\times10^{-3}$ | $145.157\times10^{-3}$ | $157.113\times10^{-3}$ |
| $U_z$ | 0.9399 | 0.8889 | 0.8548 | 0.8429 |
| $u_e$ (kPa) | 6.01 | 11.11 | 14.52 | 15.71 |

$\sigma_1' = 29.7 + 100.0 = 129.7$ kPa. Using Equation 6.3:

$$S_c = \frac{0.5\times3.0}{1+1.3}\log\frac{129.7}{29.7} = 0.417 \text{ m}.$$

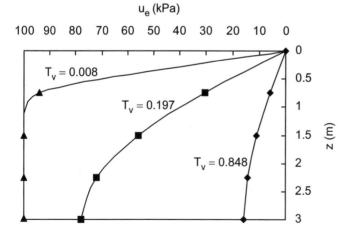

Figure 6.3. Problem 6.7.

Follow the step procedure explained in Problem 6.2.
From Equation 6.6:

$$T_v = \frac{6.5 \times 2.0}{3.0^2} = 1.444.$$

The average degree of consolidation is computed from Equations 6.10:

$$T_v = 1.444 = -0.933 \log(1-U) - 0.085,$$

$U = 0.9770$, thus

$$U = 0.9770 = \frac{S}{0.417} \rightarrow S = 0.407 \text{ m}.$$

Problem 6.8

If the water level in Problem 6.7 is lowered to the surface of the clay layer; calculate the consolidation settlement of the clay layer after 6 months if:
(a) the drawdown is instantaneous,
(b) the drawdown takes 2 months.
Assume there is no surface load and take the unit weight of the sand layer 17.5 kN/m$^3$ after the drawdown has taken place.

Solution:

(a) From Problem 6.7: $\sigma_0'$ (at the centre of clay layer) = 29.7 kPa.

$\sigma_1'$ (at the centre of clay layer) = $16.0 \times 1.5 + 17.5 \times 2.0 - 1.5 \times 9.81 = 44.3$ kPa.

Using Equation 6.3:

$$S_c = \frac{0.5 \times 3.0}{1+1.3} \log \frac{44.3}{29.7} = 0.113 \text{ m}.$$

From Equation 6.6:

$$T_v = \frac{6.5 \times (6.0/12)}{3.0^2} = 0.361.$$

$$Tv = 0.361 = -0.933 \log(1-U) - 0.085,$$

$U = 0.6674$, thus

$$U = 0.6674 = \frac{S}{S_c} = \frac{S}{0.113},$$

$S \approx 0.075$ m.
(b) Terzaghi's method for the correction of the time-settlement relationship is based on the following assumptions. The corrected settlement at $t \geq t_c$ (in this example $t_c = 2$ months) is equal to the settlement on the uncorrected curve corresponding to $t / 2$. For $t < t_c$ further correction is needed by considering the load ratio at the time of interest. The load ratio is the ratio of the load at the time $t$ to the final load $\Delta\sigma$. The settlement corresponding to $t / 2$ is multiplied by this ratio. This method implies that for a specified $U$ the amount of time is twice of the time required for instantaneous loading.

As $t \geq t_c$ therefore from Equation 6.6:

$$T_v = \frac{6.5 \times (6.0/2)/12}{3.0^2} = 0.1805.$$

$T_v = \frac{\pi}{4} U^2 = 0.1805 \rightarrow U = 0.479 < 0.6$. Thus

$$U = 0.479 = \frac{S}{0.113} \rightarrow S \approx 0.054 \text{ m}.$$

Problem 6.9

A stratum of clay is 5 m thick and is overlain by 3 m of sand, the top of which is the ground surface that is subjected to a widespread load of 200 kPa. The water table is 1.5 m below the ground surface, and the pore pressure at the impermeable boundary was measured to be 242.5 kPa after 18 months. If the settlement of the ground surface was 230 mm, determine the field values of $c_v$ and $C_c$, the final consolidation settlement, and the settlement and pore pressure at the base of the clay layer after 3 years, using the concept of parabolic isochrones. The properties of the soil section are:

sand: $\rho_{dry} = 1.75$ Mg/m$^3$, $\rho_{sat} = 2$ Mg/m$^3$,

clay: $\rho_{sat} = 1.95$ Mg/m$^3$, $e_0 = 0.8$.

Solution:

Two general types of the parabolic isochrones are shown in Figure 6.4. The middle portion of the isochrone corresponding to a low value of $T_v$ (the left hand parabola) is approximately a vertical line, indicating high excess pore pressures and that the process of consolidation has not yet progressed into this part. It can be shown that the time factor $T_c$ corresponding to $z_0 = d$, is 1/12. The isochrone with a time factor of $T_v > 1/12$ represents a more developed case and is well approximated by a parabola. For a uniform distribution of the initial excess pore pressure and $T_v < 1/12$ the excess pore pressure, average degree of consolidation and $z_0$ are expressed by:

$$u_e = u_i \left( \frac{-z^2}{12 c_v t} + \frac{2z}{\sqrt{12 c_v t}} \right) \qquad (6.14)$$

$$U = 2 \sqrt{\frac{T_v}{3}} \qquad (6.15)$$

$$z_0 = \sqrt{12 c_v t} \qquad (6.16)$$

For $T_v > 1/12$:

$$u_e = u_i \left( -\frac{z^2}{d^2} + \frac{2z}{d} \right) \exp \left( \frac{1}{4} - \frac{3 c_v t}{d^2} \right) \qquad (6.17)$$

$$U = 1 - \frac{2}{3} \exp \left( \frac{1}{4} - 3 T_v \right) \qquad (6.18)$$

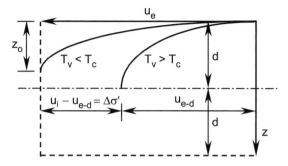

Figure 6.4. Parabolic isochrones.

The excess pore pressure on the impermeable boundary is:

$u_e = 242.5 - (5.0 + 1.5)9.81 = 178.7$ kPa.

Thus assuming $T_v > 1/12$ and using Equation 6.17:

$$u_e = 178.7 = 200.0 \left( -\frac{5.0^2}{5.0^2} + \frac{2 \times 5.0}{5.0} \right) \exp\left( \frac{1}{4} - \frac{3c_v \times 18.0/12}{5.0^2} \right),$$

$$178.7 = 200.0 \exp\left( \frac{1}{4} - \frac{4.5c_v}{25.0} \right),$$

$c_v = 2.0 \text{ m}^2/\text{year}.$

Check $T_v$ from Equation 6.6:

$$T_v = \frac{2.0 \times 18.0/12}{5.0^2} = 0.12 > 1/12 = 0.083.$$

The area between the left hand side of the isochrone (plotted in the $u_e$-$z$ coordinate system) and $z$-axis (area $S_1$ in Figure 6.5), when multiplied by $m_v$, will yield the settlement at that specific time. Similarly, the total area $(z \times u_i)$ multiplied by $m_v$ will equal the final

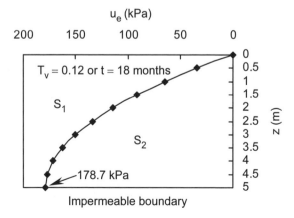

Figure 6.5. Problem 6.9: parabolic isochrone for $T_v = 0.12$ or $t = 18$ months.

settlement providing that $m_v$ is constant within the layer:

$$S_c = m_v \times (d \text{ or } 2d) \times u_i - m_v \times (d \text{ or } 2d) \times \frac{2}{3} u_{e-d} \qquad (6.19)$$

where $u_{e-d}$ is the excess pore pressure on the impermeable boundary with $d$ representing the thickness of layer. In the open layer the thickness of the layer is represent by $2d$ and $u_{e-d}$ is the excess pore pressure at the mid-point of the layer. Therefore:

$$S_c = 0.23 = m_v \times 5.0 \times 200.0 - m_v \times 5.0 \times \frac{2}{3} \times 178.7,$$

$$m_v = 5.69 \times 10^{-4} \ m^2/kN.$$

From Equation 6.4:

$$m_v = 5.69 \times 10^{-4} = \frac{\Delta e/(1+e_0)}{\sigma_1' - \sigma_0'} = \frac{\Delta e/(1.0+0.8)}{\sigma_1' - \sigma_0'}.$$

where $\sigma_1'$ and $\sigma_0'$ are initial and final (or at the end of a specified time) effective vertical stresses at the centre of the clay layer.

$$\sigma_0' = 1.95 \times 9.81 \times 2.5 + 2.0 \times 9.81 \times 1.5 + 1.75 \times 9.81 \times 1.5 - 1.0 \times 9.81(2.5 + 1.5),$$

$$\sigma_0' = 63.8 \ kPa.$$

$$\sigma_1' \ (\text{final}) = 63.76 + 200.0 = 263.8 \ kPa.$$

Substitute $\sigma_1'$ and $\sigma_0'$ in the $m_v$ equation:

$$m_v = 5.69 \times 10^{-4} = \frac{\Delta e/1.8}{200.0},$$

$$\Delta e = 0.205.$$

From Equation 6.1:

$$C_c = \frac{\Delta e}{\log(\sigma_1'/\sigma_0')} = \frac{0.205}{\log(263.8/63.8)} = 0.332.$$

To find the final settlement we use Equation 6.11:

$$U = \frac{S}{S_c} = \frac{0.23}{S_c}.$$

where $U$ is the average degree of consolidation of the layer after 18 months ($T_v = 0.12$). Using Equation 6.18:

$$U = 1 - \frac{2}{3} \exp\left(\frac{1}{4} - 3 \times 0.12\right) = 0.4028. \text{ Thus}$$

$$U = 0.4028 = \frac{0.23}{S_c} \rightarrow S_c = 0.57 \ m.$$

To compute the settlement after 3 years we follow the step procedure of Problem 6.2:

$$T_v = \frac{2.0 \times 3.0}{5.0^2} = 0.24,$$

$$U = 1 - \frac{2}{3}\exp\left(\frac{1}{4} - 3 \times 0.24\right) = 0.5833,$$

$U = S / 57.0 = 0.5833 \rightarrow S = 0.33$ m.

Excess pore pressure at the base of the clay layer is calculated from Equation 6.17:

$$u_e = 200.0\left(-\frac{5.0^2}{5.0^2} + \frac{2 \times 5.0}{5.0}\right)\exp\left(\frac{1}{4} - \frac{3 \times 2.0 \times 3.0}{5.0^2}\right) = 125.0 \text{ kPa},$$

$u$ (after 3 years) $= 125.0 + (5.0 + 1.5) \times 9.81 = 188.8$ kPa.
One may estimate the pore pressure by adding the average excess pore pressure within the layer to the hydrostatic pore pressure that existed before loading:
$u$ (after 3 years) $= 200.0(1 - 0.5833) + (5.0 + 1.5) \times 9.81 \approx 147$ kPa.

Problem 6.10

In the soil profile of Problem 6.9, vertical drains of diameter 0.3 m are constructed in a square pattern. It is required that 95% of the combined consolidation be achieved after 1.5 years. Calculate the required distance between the vertical drains. $c_h = 4$ m$^2$/year.

Solution:

Vertical drains are arranged in triangular or square patterns. The influence zone of each drain defines a soil cylinder of diameter $D_e$ that is a function of the distance $L$ between the drains:

$$D_e = 1.128\ L \quad \text{rectangular pattern,} \quad D_e = 1.505\ L \quad \text{triangular pattern} \tag{6.20}$$

The parameters $c_h$ (the coefficient of consolidation in the horizontal) and $U_h$ (the average degree of consolidation in the horizontal) are defined by:

$$c_h = \frac{k_h}{m_v \gamma_w} \tag{6.21}$$

$$U_h = 1 - \exp\left[\frac{-8T_h}{F(n)}\right] \tag{6.22}$$

where $k_h$ is the coefficient of permeability in the horizontal direction and $n$, $F(n)$, and $T_h$ are defined below in which $D_w$ is the diameter of the drain:

$$n = \frac{D_e}{D_w} \tag{6.23}$$

$$T_h = \frac{c_h t}{D_e^2} \tag{6.24}$$

$$F(n) = \frac{n^2}{n^2 - 1}\ln n - \frac{3n^2 - 1}{4n^2} \tag{6.25}$$

For the values of $n > 10$, $F(n)$ may be approximated by:

$$F(n) = \ln n - \frac{3}{4} \tag{6.26}$$

The combined excess pore pressure is given by:

$$u_e = \frac{u_{ev} \times u_{eh}}{u_i} \tag{6.27}$$

where $u_{ev}$ is the excess pore pressure due to vertical drainage only (Terzaghi's solution) and $u_{eh}$ is the excess pore pressure due to radial flow. Consequently, the combined degree of consolidation can be written in the following form:

$$U = 1 - (1 - U_h)(1 - U_v) \tag{6.28}$$

where $U_h$ and $U_v$ are the degrees of consolidation for radial and vertical flow respectively. From Problem 6.9 for 1.5 years $T_v = 0.12$ and $U_v = 0.4028$. Using Equation 6.28:

$$U = 0.95 \text{ (given data)} = 1 - (1 - U_h)(1 - 0.4028) \rightarrow U_h = 0.9163.$$

From Equations 6.23 and 6.24:

$$n = \frac{D_e}{D_w} = \frac{1.128L}{0.3} = 3.76L, \text{ therefore:}$$

$$T_h = \frac{4.0 \times 1.5}{(1.128L)^2} = \frac{4.716}{L^2}.$$

Substituting $T_h$ and $n$ values in Equation 6.22:

$$U_h = 0.9163 = 1 - \exp\left[-8 \times \frac{4.716}{L^2 \times F(n)}\right].$$

The value of $F(n)$ is substituted from Equation 6.26:

$$U_h = 0.9163 = 1 - \exp\left[-8 \times \frac{4.716}{L^2 \times (\ln n - 3/4)}\right],$$

$$U_h = 0.9163 = 1 - \exp\left[-8 \times \frac{4.716}{L^2 \times (\ln 3.76L - 3/4)}\right].$$

The above equation may be solved by trial and error from which $L = 3.0$ m will reasonably satisfy the equation.

Note: $n = 3.76 L = 11.28 > 10$, which justifies the use of Equation 6.26.

Problem 6.11

In Problem 6.7, calculate the consolidation settlements at 3 months, 6 months and 2 years if the load is increased linearly to 100 kPa over 6 months.

Solution:

For $t = 3$ months:

$$T_v = \frac{c_v t}{d^2} = \frac{6.5 \times (3.0/2)/12.0}{3.0^2} = 0.0903,$$

$$T_v = \frac{\pi U^2}{4} = 0.0903 \rightarrow U = 0.3391.$$

Correction for linear loading:
$$U = 0.3391 \times 3.0 / 6.0 = 0.1695.$$
$$S = S_c \times 0.1695 = 0.417 \times 0.1695 = 0.0707 \text{ m} \approx 71 \text{ mm}.$$

For $t = 6$ months:

$$T_v = \frac{c_v t}{d^2} = \frac{6.5 \times (6.0/2)/12.0}{3.0^2} = 0.1805,$$

$$T_v = \frac{\pi U^2}{4} = 0.1805 \rightarrow U = 0.4794.$$

No further correction is needed, thus
$$S = S_c \times 0.4794 = 0.417 \times 0.4794 = 0.1999 \text{ m} \approx 200 \text{ mm}.$$

For $t = 2$ years:

$$T_v = \frac{c_v t}{d^2} = \frac{6.5 \times (2.0/2)}{3.0^2} = 0.7222,$$

$$T_v = -0.933 \log(1-U) - 0.085 = 0.7222 \rightarrow U = 0.8636 > 0.6.$$

No further correction is needed, thus
$$S = S_c \times 0.8636 = 0.417 \times 0.8636 = 0.3601 \text{ m} \approx 360 \text{ mm}.$$

## 6.3   REFERENCES AND RECOMMENDED READINGS

Al-Khafaji, A.W. & Andersland, O.B. 1992. Equations for compression index approximations. *Journal GED, ASCE*, 118(1): 148-153.

ASTM D-2435. 1996. *Standard test method for one-dimensional consolidation properties of soils.* PA, West Conshohocken: American Society for Testing and Materials.

Australian Standard AS 1289.6.6.1. 1998. *Methods of testing for engineering purposes, method 6.6.1: soil strength and consolidation tests-determination of one-dimensional consolidation properties of a soil-standard method.* Australia, NSW: Standard Association of Australia.

Aysen, A. 2002. *Soil mechanics: Basic concepts and engineering applications.* Lisse: Balkema.

Balasubramaniam, A.S. & Brenner, R.P. 1981. Consolidation and settlement of soft clay. In E.W. Brand & R.P. Brenner (eds), *Soft clay engineering: Developments in geotechnical engineering 20.* New York: Elsevier Scientific Publishing Company.

Barron, R.A. 1948. Consolidation of fine-grained soil by drain wells. *Transactions, ASCE*, 113: 718-742.

Bergado, D.T., Anderson, L.R., Miura, A.S. & Balasubramaniam, A.S. 1996. *Soft ground improvement in lowland and other environments.* USA: ASCE Press.

BS 1377-6. 1990. *Methods of test for soils for civil engineering purposes. Consolidation and permeability tests in hydraulic cells and with pore pressure measurement.* London: British Standards Institution.

Casagrande, A. 1936. Discussion 34: The determination of preconsolidation load and its practical significance. *Proc. 1st intern. conf. SMFE*, 3: 60-64. Cambridge.

Craig, R.F. 1997. *Soil mechanics.* 6th edition. London: E & FN Spon.

Duncan, J.M. 1993. Limitations of conventional analysis of consolidation settlement. *Journal GE, ASCE,* 119(9): 1333-1359.

Hansbo, S. 1981. Consolidation of fine-grained soils by prefabricated drains. *Proc. 10th intern. conf. SMFE,* 3: 12-22. Stockholm.

Hansbo, S. 1987. Design aspects of consolidation of vertical drains and lime column installation. *Proc. 9th Southeast Asian geotechnical con.* 2: 8-12. Bangkok, Thailand.

Koppula, S.D. 1986. Discussion: consolidation parameters derived from index tests. *Geotechnique,* 36(2): 68-73.

Leonards, G.A. 1976. Estimating consolidation settlements of shallow foundations on overconsolidated clays. *Special report 163, transportation research board*: 13-16.

Mesri, G. & Castro, A. 1987. The $C_\alpha / C_c$ concept and $k_o$ during secondary compression. *Journal GE, ASCE,* 112(3): 230-247.

Mesri, G. & Godlewski, P.M. 1977. Time and stress compressibility interrelationship. *Journal GE, ASCE,* 103(5): 417-430.

Mesri, G. & Rokhsar, A. 1974. Theory of consolidation for clays. *Journal GED, ASCE,* 100(GT8): 889-904.

Nagaraj, T.S. & Srinivasa Murthy, B.R. 1985. Prediction of the preconsolidation pressure and recompression index of soils. *Geotechnical testing journal, ASTM,* (4): 199-202.

Olson, R.E. & Ladd, C.C. 1979. One-dimensional consolidation problems. *Journal GED, ASCE,* 105(GT1): 11-30.

Powers, J.D. 1985. Dewatering-avoiding its unwanted side effects. *Underground technology research council.* New York: ASCE.

Raymond, G.P. & Wahls, H.E. 1976. Estimating 1-dimensional consolidation, including secondary compression of clay loaded from overconsolidated to normally consolidated state. *Special report 163, transportation research board*: 17-23.

Rendon-Herrero, O. 1980. Universal compression index equation. *Journal GED, ASCE,* 106(GT11): 1179-1200.

Rendon-Herrero, O. 1983. Closure: universal compression index equation. *Journal GED, ASCE,* 109(GT5): 755-761.

Rixner, J.J., Kraemer, S.R. & Smith, A.D. 1986. Prefabricated vertical drains. *Engineering guidelines, federal highway administration,* 1(FHWA-RD-86 / 168). Washington DC.

Schmertmann, J.H. 1953. Estimating the true consolidation behavior of clay from laboratory test results. *Journal SMFE, ASCE,* 79(311): 26.

Scott, R.E. 1963. *Principles of soil mechanics.* Reading, Massachusetts: Addison-Wesley.

Skempton, A.W. 1944. Notes on the compressibility of clays. *Quart. journal of geological society,* (100): 119-135. London.

Skempton, A.W. & Bjerrum, L. 1957. A contribution to the settlement analysis of foundations on clay. *Geotechnique,* (7): 168-178.

Terzaghi, K., Peck, R. B., & Mesri, G. 1996. *Soil mechanics in engineering practice.* 3rd edition. New York: John Wiley & Sons.

CHAPTER 7

# Application of Limit Analysis to Stability Problems in Soil Mechanics

## 7.1  INTRODUCTION

The main objective of this chapter is to investigate the stability of a soil structure using the *lower* and *upper bound theorems* of plasticity. These theorems are used to predict collapse loads where analytical solutions either do not exist or are inconsistent with the governing equations of mechanics. They are also used when the deformations of the soil structure are negligible.

A lower bound solution provides a safe limit load, whereas an upper bound solution estimates an unsafe limit load under which the failure of material has taken place already. In a lower bound solution only the equilibrium and yield criterion are satisfied, whilst in an upper bound solution, only the compatibility and the yield criterion are considered. These solutions, obtained either manually or numerically, bracket the exact solution within usually acceptable accuracy (Aysen & Sloan, 1991a).

In a simplified upper bound solution (in two dimensions) the continuum is converted into a mechanism consisting of rigid blocks sliding on their contact areas. For a virtual displacement, the external work done by the external forces is equated to the internal work done by the internal forces to obtain the unknown collapse load as an upper bound to the true collapse load. The stress-strain model for the soil is idealized as a rigid-perfectly plastic material.

Problem 7.1 describes the stress discontinuity, which is the basic tool in a lower bound solution. Application of the lower and upper bounds includes bearing capacity of shallow footings (Problems 7.2 and 7.3), stability of slopes (Problems 7.4 to 7.7), retaining walls (Problems 7.8 to 7.10) and tunnel heading (Problem 7.11).

## 7.2  PROBLEMS

Problem 7.1

A stress discontinuity makes an angle of 60° with the $x$-axis (Figure 7.1). On the right-hand side of the discontinuity: $\sigma'_x = 100$ kPa, $\tau_{xz} = 50$ kPa. Determine:
(a) the magnitude of $\sigma'_z$ to satisfy the failure criterion,
(b) the normal and shear stresses on the discontinuity,
(c) the state of the stresses at points on the left-hand side of the stress discontinuity to satisfy the failure criterion. $c' = 20$ kPa, $\phi' = 30°$.

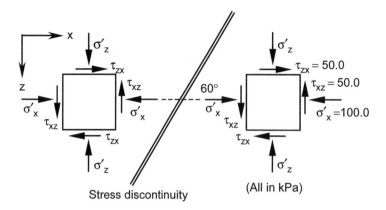

Stress discontinuity

(All in kPa)

Figure 7.1. Problem 7.1.

Solution:

(a) Substituting the given data in the failure criterion expresses by Equation 4.3:

$$(\sigma'_z - 100.0)^2 + (2 \times 50.0)^2 = [2 \times 20.0 \cos 30.0° + (\sigma'_z + 100.0) \sin 30.0°]^2,$$

$$0.75\sigma'^2_z - 284.64\sigma'_z + 12835.90 = 0,$$

$\sigma'_z = 327.2$ kPa (the higher answer is selected).

(b) The normal and shear stresses on the discontinuity are calculated from Equations 4.1 and 4.2:

$$\sigma' = \frac{327.2 + 100.0}{2} + \frac{327.2 - 100.0}{2} \cos(2 \times 60.0°) + 50.0 \sin(2 \times 60.0°) = 200.1 \text{ kPa}.$$

$$\tau = \frac{327.2 - 100.0}{2} \sin(2 \times 60.0°) - 50.0 \cos(2 \times 60.0°) = 123.4 \text{ kPa}.$$

(c) We need to create three equations in terms of $\sigma'_x$, $\sigma'_z$ and $\tau_{xz}$ of the left-hand side of the discontinuity. Two equations are obtained from equilibrium using Equations 4.1 and 4.2. The third equation is obtained by enforcing the failure criterion of Equation 4.3:

$$\sigma' = 200.1 = \frac{\sigma'_z + \sigma'_x}{2} + \frac{\sigma'_z - \sigma'_x}{2} \cos 120.0° + \tau_{xz} \sin 120.0°,$$

$$\tau = 123.4 = \frac{\sigma'_z - \sigma'_x}{2} \sin 120.0° - \tau_{xz} \cos 120.0°.$$

Combining the above two equations we obtain:

$$(\sigma'_z + \sigma'_x) = 2(\sigma'_z - \sigma'_x) - 27.270.$$

The second equation may be written in the following form:

$$\tau_{xz} = -0.866(\sigma'_z - \sigma'_x) + 246.8. \text{ Thus}$$

$$(\sigma'_z - \sigma'_x)^2 + 2^2[-0.866(\sigma'_z - \sigma'_x) + 246.8]^2 =$$

$$\{2 \times 20.0 \cos 30.0° + [2 \times (\sigma'_z - \sigma'_x) - 27.270] \sin 30.0°\}^2.$$

The solution of the above equation yields:

$\sigma'_z - \sigma'_x = 356.97$ kPa (the higher answer is selected).

$\tau_{xz} = -0.866(\sigma'_z - \sigma'_x) + 246.8 = -0.866 \times 356.97 + 246.8 = -62.3$ kPa.

$(\sigma'_z + \sigma'_x) = 2(\sigma'_z - \sigma'_x) - 27.270 = 2 \times 356.97 - 27.27 = 686.67$ kPa.

Solving the following two equations:

$\sigma'_z + \sigma'_x = 686.67$ kPa,

$\sigma'_z - \sigma'_x = 356.97$ kPa.

$\sigma'_z = 521.8$ kPa, $\sigma'_x = 164.8$ kPa.

Problem 7.2

Figure 7.2 shows a strip footing on $c_u$, $\phi_u = 0$ with 3 stress discontinuities. It is required to calculate the ultimate bearing capacity ($q_u$) of the footing and the stress field within each zone. Assume that the footing is smooth and take into account the weight of the material.

Solution:

The stress field within each zone along the depth $z$ (Figure 7.2) is constant and $F = 0$, therefore: $\sigma_1 - \sigma_3 = 2c_u$.

The state of stress in zone 1 and at depth $z$ is:

$\sigma_3 = \gamma z, \sigma_1 = 2c_u + \gamma z$.

The angles $\alpha_1$, $\alpha_2$, $\alpha_3$, and $\alpha_4$ (all unknown) are selected in a way that the rotation of the principal stress directions across each discontinuity are all equal and have the identical value of $90.0° / 3 = 30.0°$, where $90.0°$ is the total rotation (zone 1 to zone 4).

The angles of the major principal planes in zones 1 and 2 with the stress discontinuity $CA$ are $\phi_1$ and $\phi_2$ respectively.

For $c_u$, $\phi_u = 0$ soil:

$$\phi_1 + \phi_2 = 90° \tag{7.1}$$

Rotation of the principal stress directions due to the stress discontinuity is defined by $\psi$ that is given by:

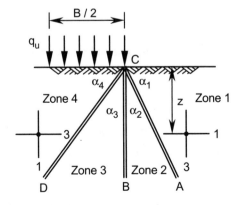

Figure 7.2. Problem 7.2.

$\psi = \varphi_2 - \varphi_1 = 30.0°$ and from Equation 7.1:

$\varphi_2 + \varphi_1 = 90.0°$. Solving for $\varphi_1$ and $\varphi_2$: $\varphi_1 = 30.0°$, $\varphi_2 = 60.0°$.

The normal and shear stresses on the stress discontinuity $CA$ are calculated from the state of stress in zone 1:

$$\sigma_{CA} = \frac{\sigma_1 + \sigma_3}{2} + \frac{\sigma_1 - \sigma_3}{2} \cos 2\varphi_1,$$

$$\sigma_{CA} = \frac{2c_u + \gamma z + \gamma z}{2} + \frac{2c_u + \gamma z - \gamma z}{2} \cos(2 \times 30.0°),$$

$$\sigma_{CA} = 1.5 c_u + \gamma z.$$

$$\tau_{CA} = \frac{\sigma_1 - \sigma_3}{2} \sin 2\varphi_1,$$

$$\tau_{CA} = \frac{2c_u + \gamma z - \gamma z}{2} \sin(2 \times 30.0°) = 0.866 c_u.$$

Note that: $\alpha_1 = 90.0° - \phi_1 = 90.0° - 30.0° = 60.0°$.

To calculate the principal stresses in zone 2, the normal and shear stresses on the stress discontinuity $CA$ are expressed in terms of the principal stresses in zone 2:

$$\sigma_{CA} = 1.5 c_u + \gamma z = \frac{\sigma_1 + \sigma_3}{2} + \frac{\sigma_1 - \sigma_3}{2} \cos 2\varphi_2,$$

$$\tau_{CA} = 0.866 c_u = \frac{\sigma_1 - \sigma_3}{2} \sin 2\varphi_2.$$

Noting that $\sigma_1 - \sigma_3 = 2c_u$, and $\varphi_2 = 60.0°$, the stress state at zone 2 is obtained by solving the above two equations:

$$\sigma_3 = c_u + \gamma z, \sigma_1 = 3c_u + \gamma z.$$

Performing a similar calculation and relating the states of stress between zones 2 and 3 and between zones 3 and 4 we find:

Zone 3:

$\alpha_2 = 30.0°, \sigma_3 = 2c_u + \gamma z, \sigma_1 = 4c_u + \gamma z,$

Zone 4:

$\alpha_3 = 30.0°, \sigma_3 = 3c_u + \gamma z, \sigma_1 = 5c_u + \gamma z,$ and $\alpha_4 = 60.0°$.

Under the smooth footing $z = 0$, thus $q_u = \sigma_1 = 5c_u$.

The magnitude of $q_u$ is independent from the unit weight of the soil and is less than the analytical value of $5.14 c_u$.


Problem 7.3

Calculate the bearing capacity factor $N_c$ for undrained conditions for a smooth strip footing if the ground surface outside of the foundation makes an angle of 15° above or below the horizontal foundation level.

Solution:

Based on the solution presented in Problem 7.2, a general equation for $N_c$ in undrained conditions can be obtained:

$$q_u = 2(c_u + nc_u \sin \psi) = 2c_u(1 + n\sin \psi),$$

$$N_c = 2(1 + n\sin \psi) \tag{7.2}$$

where $n$ is the total number of stress discontinuities, and $\psi$ is the rotation of the principal stress directions per one stress discontinuity. When the ground surface outside of the foundation makes an angle of $15°$ above the horizontal foundation level, the total rotation of the principal stresses is:

$\psi$ (total) $= 90.0° + 15.0° = 105.0°$.

Note that only half of the section (similar to Figure 7.2) is taken into account. For example, for 2 stress discontinuity (3 stress zones):

$$N_c = 2\left(1 + 2 \times \sin \frac{105.0°}{2}\right) = 5.17.$$

For $n = 100$:

$$N_c = 2\left(1 + 100 \times \sin \frac{105.0°}{100}\right) = 5.66.$$

When the ground surface outside of the foundation makes an angle of $15°$ below the horizontal foundation level:

$\psi$ (total) $= 90.0° - 15.0° = 75.0°$. For 2 stress discontinuity (3 stress zones):

$$N_c = 2\left(1 + 2 \times \sin \frac{75.0°}{2}\right) = 4.43.$$

For $n = 100$: $N_c = 2\left(1 + 100 \times \sin \frac{75.0°}{100}\right) = 4.62.$

## Problem 7.4

For the plane strain slope of $45°$ shown in Figure 7.3(a), calculate the lower and upper bound values for $q / c_u$. For the lower bound solution select a reasonable value for the number of the stress discontinuities passing through point $d$ and apply the concept used in Equation 7.2. For the upper bound solution use the mechanism shown in Figure 7.3(a). Assume the stability number $\gamma H / c_u = 0$, $c_u = 100$ kPa.

Solution:

In the zone bounded by line $ab$, $\sigma_1$ is vertical whilst in the zone bounded by $cd$ it is parallel to $cd$; thus $\sigma_1$ has rotated $45.0°$: $\psi$ (total) $= 45.0°$.
For $n = 100$ stress discontinuity:

$$\frac{q}{c_u} = N_c = 2\left(1 + 100 \times \sin \frac{45.0°}{100}\right) = 3.57.$$

Note that by using the Bishop's simplified slice method (Chapter 9) a slightly higher answer of 3.68 will be achieved. From the geometry of the problem:

$h_1 = 17.32$ m, $h_2 = 12.68$ m, $l = 40.0$ m, $\alpha_1 = 17.6°$, $\delta = 27.4°$,

$ab = bd = 20.0$ m, $bc = 41.96$ m.

The displacement diagram of the sliding blocks are shown in Figure 7.3(b), where $v_1$ and $v_2$ are the displacements of blocks 1 (*abd*) and 2 (*bcd*) respectively and $v_{12}$ is the relative displacement between two blocks.

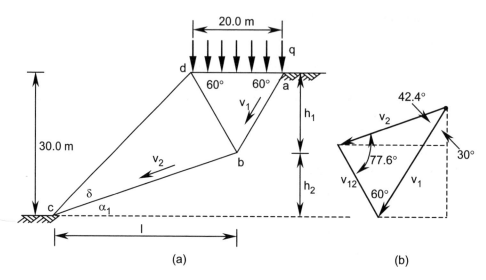

Figure 7.3. Problem 7.4.

From the geometry of the displacement diagram and assuming $v_1 = 1$, we find $v_2 = 0.887$ and $v_{12} = 0.690$. The external work has one component due to the surface load $q$, only, because the material is assumed weightless:

$E_e = 20.0 \times q \times v_1 \times \cos 30.0° = 17.32q$.

The flow rule in a rigid-perfectly plastic material requires the displacement to make an angle of $\phi'$ or $\phi_u$ with the sliding surface. It can be shown that the internal works are due only to cohesion and must have identical signs (positive preferred); thus

$E_{i1} = ab \times c_u \times v_1 \cos\phi_u = 20.0 \times 100.0 \times 1.0 = 2000.00$,

$E_{i2} = bc \times c_u \times v_2 \cos\phi_u = 41.96 \times 100.0 \times 0.887 = 3721.85$,

$E_{i3} = bd \times c_u \times v_{12} \cos\phi_u = 20.0 \times 100.0 \times 0.690 = 1380.00$.

Total internal work = 7101.85. Equating the external work and the internal work:
$17.32q = 7101.85$, $q = 410.0$ kPa, $q / c_u = 410.0 / 100.0 = 4.1$.

Note that the given angles in the geometry of the mechanism are not optimized values. The exact value of $q / c_u$ is between 3.57 and 4.1.

Problem 7.5

A plane strain vertical cut is subjected to a vertical uniform load $q$ at the upper ground surface (Figure 7.4(a)). If the lower bound for the stability number $\gamma H / c'$ (with $q / c' = 0$) is 6.12, (this is obtained from a finite element analysis) calculate a lower bound for the load parameter $q / c'$ for the given soil parameters.

$c' = 20$ kPa, $\phi' = 30°$, $\gamma = 19$ kN/m$^3$, and $H = 5$ m.

Solution:

A statically admissible stress field with 2 stress discontinuities (3 stress zones) is shown in Figure 7.4(a). The stress field (either effective or total) in the three zones above may be summarized as follows:

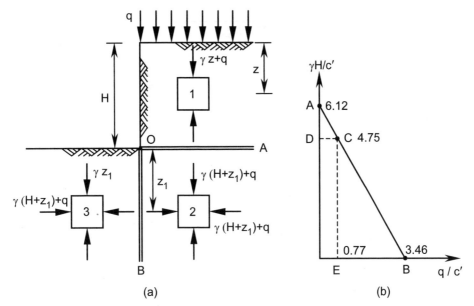

Figure 7.4. Problem 7.5.

Zone 1: $\sigma'_3 = 0, \sigma'_1 = \gamma z + q$. Zone 2: $\sigma'_3 = \gamma(H + z_1) + q, \sigma'_1 = \gamma(H + z_1) + q$.
Zone 3: $\sigma'_3 = \gamma z_1, \sigma'_1 = \gamma(H + z_1) + q$.

For a $c'$, $\phi'$ material, failure occurs only at the toe of the slope. The relationship between $\sigma'_1$ and $\sigma'_3$ at the toe is given by Equation 4.8 (or 4.9). Replacing $\sigma'_3$ and $\sigma'_1$ by the corresponding values defined above, two identical equations in terms of the stability number and the load parameter are obtained.

$$\frac{\gamma H}{c'} = 2\tan(45° + \phi'/2) \to q = 0 \qquad (7.3)$$

$$\frac{q}{c'} = 2\tan(45° + \phi'/2) \to \frac{\gamma H}{c'} = 0 \qquad (7.4)$$

The highest lower bounds found for the stability number $\gamma H / c'$ and load parameter $q / c'$ may be related by a linear relationship, as shown by the line $AB$ in Figure 7.4(b). This allows the estimation of a lower bound for one of the parameters when the magnitude of the other parameter is known. Thus from Equation 7.4:

$$\frac{q}{c'}\left(\text{for } \frac{\gamma H}{c'} = 0\right) = 2\tan(45.0° + 30.0°/2) = 3.46.$$

For the case $q = 0$, a higher value of 6.12 is given; therefore Equation 7.3 will not be used. The results are shown in Figure 7.4(b) from which the equation of line $AB$ becomes:

$\dfrac{q}{c'} = -0.5654\dfrac{\gamma H}{c'} + 3.46$. The stability number corresponding to the given data is:

$\dfrac{\gamma H}{c'} = \dfrac{19.0 \times 5.0}{20.0} = 4.75$. Substituting this value in the equation of line $AB$:

$\dfrac{q}{c'} = -0.5654 \times 4.75 + 3.46 = 0.774 \to q = 0.774 \times 20.0 = 15.5$ kPa.

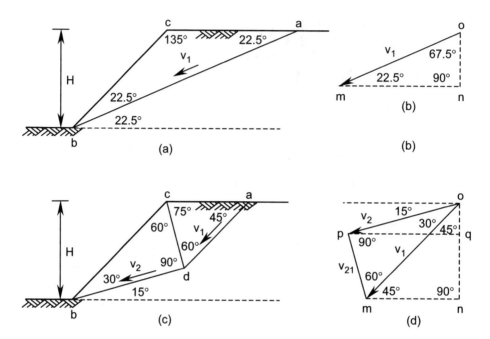

Figure 7.5. Problem 7.6: (a) & (b) single mechanism, (c) & (d) two-block mechanism.

## Problem 7.6

For a 45° plane strain slope in a $c_u$, $\phi_u = 0$ soil, compute the upper bound values for the stability number $\gamma H / c_u$ using the collapse mechanism shown in:
(a) Figure 7.5(a) for a single block and
(b) Figure 7.5(c) for two blocks.

Solution:

(a) The displacement diagram is shown in Figure 7.5(b), where $v_1$ (or *om*) represents the displacement across the slip plane and *on* is the vertical displacement of the sliding block. From the geometry of the mechanism and displacement diagram we have:

$$bc = ca = 1.414H, ab = 2.613H, on = 0.383v_1 \rightarrow v_1 = 1.0 \rightarrow on = 0.383.$$

$w = $ weight of the block $abc = \dfrac{1}{2}H \times 1.414H \times 1.0 \times \gamma = 0.707\gamma H^2$.

Compute external work:

$$E_e = w \times on = 0.707\gamma H^2 \times 0.383 = 0.271\gamma H^2.$$

Compute internal work:

$$E_i = ab \times c_u \times v_1 \cos\phi_u = 2.613H \times c_u \times 1.0 = 2.613Hc_u.$$

Equate external work to internal work:

$$0.271\gamma H^2 = 2.613Hc_u,$$

$$\frac{\gamma H}{c_u} = 9.64.$$

(b) The displacement diagram is shown in Figure 7.5(d), where $v_1$ (or *om*) and $v_2$ (or *op*) represent the displacements of blocks across the two slip planes and *on* and *oq* are the vertical displacements of the sliding blocks. The displacement $v_{21}$ is the relative displacement of block 2 against block 1 on the corresponding slip plane (*dc* in Figure 7.5(c)). From the geometry of the mechanism and displacement diagram we have:

$$bc = 1.414H, cd = 0.707H, bd = 1.225H, da = 0.966H,$$

$$v_1 = 1.0, v_2 = 0.866, v_{21} = 0.5, on = 0.707, oq = 0.224.$$

$w_1 =$ weight of the block $1 = \dfrac{1}{2}0.707H\sin 60.0° \times 0.966H \times 1.0 \times \gamma = 0.296\gamma H^2.$

$w_2 =$ weight of the block $2 = \dfrac{1}{2}1.225H \times 0.707H \times 1.0 \times \gamma = 0.433\gamma H^2.$

Compute external work:

$$E_e = w_1 \times on + w_2 \times oq = 0.296\gamma H^2 \times 0.707 + 0.433\gamma H^2 \times 0.224 = 0.306\gamma H^2.$$

Compute internal work:

$$E_i = ad \times c_u \times v_1 \cos\phi_u + db \times c_u \times v_2 \cos\phi_u + cd \times c_u \times v_{21}\cos\phi_u \rightarrow \cos\phi_u = 1.0,$$

$$E_i = 0.966H \times c_u \times 1.0 + 1.225H \times c_u \times 0.866 + 0.707H \times c_u \times 0.5 = 2.380Hc_u.$$

Equate external work to internal work:

$$0.306\gamma H^2 = 2.380Hc_u \rightarrow \frac{\gamma H}{c_u} = 7.78 < 9.64 \text{ (single mechanism)}.$$

## Problem 7.7

For a 45° plane strain slope in a $c_u$, $\phi_u = 0$ soil, calculate the load parameter $q / c_u$ using the collapse mechanism of Figure 7.6(a). $\gamma H / c_u = 0$.

Solution:

From the geometry of the mechanism and displacement diagram:

$$bc = 1.414H, bd = 1.454H, da = ca = cd = 0.863H,$$

$$v_1 = 1.0, v_2 = 0.922, v_{21} = 0.815, on = 0.866.$$

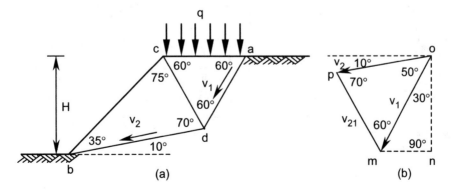

Figure 7.6. Problem 7.7.

Compute external work:

$E_e = q \times ca \times on = q \times 0.863H \times 0.866 = 0.747qH$.

Compute internal work:

$E_i = ad \times c_u \times v_1 \cos \phi_u + db \times c_u \times v_2 \cos \phi_u + cd \times c_u \times v_{21} \cos \phi_u \rightarrow \cos \phi_u = 1.0$,

$E_i = 0.863H \times c_u \times 1.0 + 1.454H \times c_u \times 0.922 + 0.863H \times c_u \times 0.815 = 2.907Hc_u$.

Equate external work to internal work:

$0.747 \gamma H^2 = 2.907 Hc_u \rightarrow \dfrac{\gamma H}{c_u} = 3.89 < 4.10$ (Problem 7.4).

## Problem 7.8

A 10 m height of saturated clay is supported by a rough retaining wall. The properties of the soil are: $c_u = 60$ kPa, $\phi_u = 0$, and $\gamma = 18$ kN/m$^3$. The vertical boundary load is $q = 50$ kPa and $c_w$ (cohesion mobilized between wall and soil) = 30 kPa. Calculate the lower bound value of the horizontal active thrust and the position of its point of application.

Solution:

The formulation for $c_u$, $\phi_u = 0$ soil and a rough wall with one stress discontinuity (Figure 7.7(a)) is as follows:

The rotation of the principal stresses along the stress discontinuity is:

$$\psi = \frac{\theta}{2} \tag{7.5}$$

where $\theta$ is defined by:

$$\sin \theta = \frac{c_w}{c_u} \tag{7.6}$$

and the angle $\alpha$ is found to be:

$$\alpha = \frac{\pi}{4} - \frac{\theta}{4} \tag{7.7}$$

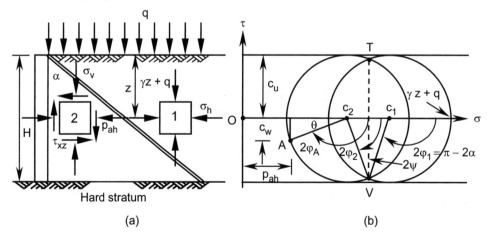

Figure 7.7. Problem 7.8.

From the geometry of Figure 7.7(b) the active lateral pressure $p_{ah}$ at depth $z$ is:

$$p_{ah} = (\gamma z + q) - c_u\left(1 + 2\sin\frac{\theta}{2} + \cos\theta\right)$$
(7.8)

Integrating the above equation along the wall, the total horizontal active thrust is expressed by:

$$P_{ah} = \frac{\gamma H^2}{2} + \left[q - c_u\left(1 + 2\sin\frac{\theta}{2} + \cos\theta\right)\right]H$$
(7.9)

If the number of stress discontinuities is increased from 1 to $n$, then the term $2\sin(\theta/2)$ in Equations 7.8 and 7.9 is replaced by $2n\sin(\theta/2n)$. However, the improvement in the active thrust is not significant. The passive pressure can be calculated in a similar manner:

$$P_{ph} = (\gamma z + q) + c_u\left(1 + 2\sin\frac{\theta}{2} + \cos\theta\right)$$
(7.10)

The total horizontal passive thrust is:

$$P_{ph} = \frac{\gamma H^2}{2} + \left[q + c_u\left(1 + 2\sin\frac{\theta}{2} + \cos\theta\right)\right]H$$
(7.11)

Note that if the horizontal active pressure (Equation 7.8) has negative value at the ground surface, Equation 7.9 cannot be used because the soil cannot sustain the tensile stress. Assuming one stress discontinuity and using Equations 7.6 and 7.8:

$$\sin\theta = \frac{c_w}{c_u} = \frac{30.0}{60.0} = 0.5 \rightarrow \theta = 30.0°.$$

$$p_{ah} = 18.0z + 50.0 - 60.0\left(1 + 2\sin\frac{30.0°}{2} + \cos 30.0°\right), p_{ah} = 18.0z - 93.02.$$

$p_{ah} = 18.0z - 93.02 = 0 \rightarrow z = z_o = 5.17$ m (depth of tension crack).

$p_{ah}(z = 10.0 \text{ m}) = 18.0 \times 10.0 - 93.02 = 86.98$ kPa.

$P_{ah}$(horizontal thrust) $= p_{ah}(z = 10.0 \text{ m}) \times (10.0 - z_o)/2$,

$P_{ah}$(horizontal thrust) $= 86.98 \times (10.0 - 5.17)/2 = 210.0$ kN.

The application point of the active thrust measured from the base =
$(10.0 - 5.17)/3 = 1.61$ m.

Problem 7.9

A retaining wall of height 6 m retains a sandy soil for which $c' = 0$ and $\phi' = 35°$, and $\gamma = 19$ kN/m³. A uniform load of $q = 50$ kPa acts on the upper ground surface. Assuming $\delta'$ (the friction angle mobilized on the interface) $= 35°$, calculate the lower and upper bounds for the active thrust (the resultant of the horizontal and vertical components). For the upper bound solution assume a single variable mechanism similar to Figure 7.9(a).

Solution:

The stress circles for a $c' = 0$, $\phi'$ material and for one stress discontinuity (similar to Figure 7.7(a)) are shown in Figure 7.8. The shear stress on an element of soil immediately behind the retaining wall is:

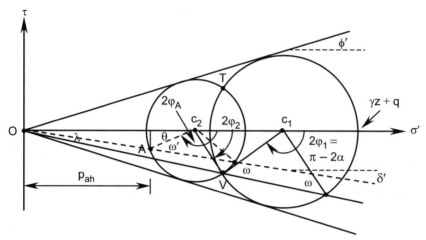

Figure 7.8. Problem 7.9.

$\tau_{xz} = p_{ah}\tan\delta'.$

The active pressure is expressed by:

$$p_{ah} = (\gamma z + q)k_{ah} = (\gamma z + q)\frac{\sin(\omega - \lambda)(1 - \sin\phi'\cos\theta)}{\sin(\omega + \lambda)(1 + \sin\phi')} \qquad (7.12)$$

where the angles $\theta$, $\omega$ and $\lambda$ are defined in Figure 7.8 and:

$$\sin\omega = \frac{\sin\lambda}{\sin\phi'} \qquad (7.13)$$

Rotation of the principal stress directions due to the stress discontinuity is defined by $\psi$ that is given by:

$$\psi = \frac{2\phi_2 - 2\phi_1}{2} = \frac{\theta}{2} = 90° - \omega \qquad (7.14)$$

The angle $\theta$ can be expressed in terms of $\delta'$ and $\phi'$ by introducing the angle $\omega'$ defined in Figure 7.8, then:

$$\sin\omega' = \sin(\theta + \delta') = \frac{\sin\delta'}{\sin\phi'} \qquad (7.15)$$

For the passive case this equation has the following form:

$$\sin\omega' = \sin(\theta - \delta') = \frac{\sin\delta'}{\sin\phi'} \qquad (7.16)$$

Using a similar approach, the soil pressure coefficient for the passive case can be obtained:

$$k_{ph} = \frac{\sin(\omega + \lambda)(1 + \sin\phi'\cos\theta)}{\sin(\omega - \lambda)(1 - \sin\phi')} \qquad (7.17)$$

The total active or passive thrusts are the integral of the soil pressure along the wall:

$$P_{ah} = \left( \frac{\gamma H^2}{2} + qH \right) k_{ah} \qquad (7.18)$$

$$P_{ph} = \left( \frac{\gamma H^2}{2} + qH \right) k_{ph} \qquad (7.19)$$

It can be shown that the increasing of number of stress discontinuities will not significantly improve the solution. Using Equation 7.15:

$\sin \omega' = \sin \delta' / \sin \phi' = \sin 35.0° / \sin 35.0° = 1.0 \to \omega' = 90.0°$,

$\omega' = 90.0° = \theta + \delta' = \theta + 35.0° \to \theta = 55.0° \to \theta / 2 = 27.5°$.

From Equation 7.14:

$\theta / 2 = 27.5° = 90.0° - \omega \to \omega = 62.5°$.

Thus the angle $\lambda$ can be found from Equation 7.13:

$$\sin \omega = \sin 62.5° = \frac{\sin \lambda}{\sin 35.0°} \to \lambda = 30.6°.$$

Calculate the horizontal component of active thrust using Equations 7.12 and 7.18:

$$k_{ah} = \frac{\sin(\omega - \lambda)(1 - \sin \phi' \cos \theta)}{\sin(\omega + \lambda)(1 + \sin \phi')} = \frac{\sin(62.5° - 30.6°)(1 - \sin 35.0° \cos 55.0°)}{\sin(62.5° + 30.6°)(1 + \sin 35.0°)} = 0.2257.$$

$P_{ah} = (\gamma H^2 / 2 + qH)k_{ah} = (19.0 \times 6.0^2 / 2 + 50.0 \times 6.0) \times 0.2257 = 144.9 \text{ kN}.$

Vertical component of active thrust is $P_{av} = 144.9 \times \tan 35.0° = 101.5 \text{ kN}$, thus

$$P_a = \sqrt{144.9^2 + 101.5^2} = 176.9 \text{ kN}.$$

From the geometry of the upper bound mechanism and the displacement diagram of Figure 7.9:

$ca = H \times \tan \alpha = 6 \tan \alpha$.

$w = H(H \times \tan \alpha / 2)\gamma = 6.0 \times 6.6 \times \tan \alpha \times 1/2 \times 1.0 \times 19.0 = 342 \tan \alpha$.

$v_a = 1.0, on = \cos(\alpha + 35.0°), nm = \sin(\alpha + 35.0°)$.

Compute external work:

$E_e = (q \times ca + w) \times on - P_a \sin 35.0° \times on - P_a \cos 35.0° \times nm$.

As $c' = 0$, therefore the internal work is zero, thus the sum of the external works must equal to zero:

$(50.0 \times 6 \tan \alpha + 342.0 \tan \alpha) \cos(\alpha + 35.0°)$

$- P_a \sin 35.0° \cos(\alpha + 35.0°) - P_a \cos 35.0° \sin(\alpha + 35.0°) = 0$.

$$P_a = \frac{642.0 \tan \alpha \times \cos(\alpha + 35.0°)}{\sin 35.0° \cos(\alpha + 35.0°) + \cos 35.0° \sin(\alpha + 35.0°)} = \frac{642.0 \tan \alpha \times \cos(\alpha + 35.0°)}{\sin(\alpha + 70.0°)}.$$

Using a trial and error method the value of $\alpha$ corresponding to the maximum value of $P_a$ is found to be 32.5°, therefore:

$$P_a = \frac{642.0 \tan 32.5° \times \cos(32.5° + 35.0°)}{\sin(32.5° + 70.0°)} = 160.3 \text{ kN}.$$

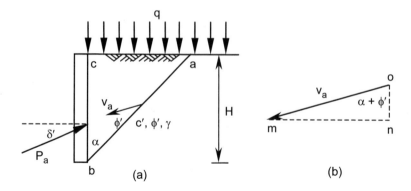

Figure 7.9. Upper bound mechanism for Problems 7.9 & 7.10.


Problem 7.10

Using the single variable mechanism of Figure 7.9(a), find the upper bound value for the active thrust using the following data:
$H = 5$ m, $\alpha = 34°$, $q = 80$ kPa, $\delta' = 20°$, $c' = 10$ kPa, $\phi' = 25°$, and $\gamma = 18$ kN/m$^3$.

Solution:

From the geometry of the upper bound mechanism and the displacement diagram of Figure 7.9(b):

$ca = 5.0 \times \tan 34.0° = 3.37$ m,

$ab = 5.0 / \cos 34.0° = 6.03$ m.

$w = 5.0 \times 3.37 \times 1.0 \times 18.0/2 = 151.7$ kN.

$v_a = 1.0, on = \cos(34.0° + 25.0°) = 0.515, nm = \sin(34.0° + 25.0°) = 0.857$.

Compute external work:

$E_e = (q \times ca + w) \times on - P_a \sin 20.0° \times on - P_a \cos 20.0° \times nm$,

$E_e = (80.0 \times 3.37 + 151.7) \times 0.515 - P_a \sin 20.0° \times 0.515 - P_a \cos 20.0° \times 0.857$,

$E_e = 217.0 - 0.981 P_a$.

Compute internal work:

$E_i = ab \times c' \times v_a \cos\phi' = 6.03 \times 10.0 \times 1.0 \times \cos 25.0° = 54.65$.

Equate external work to internal work:

$217.0 - 0.981 P_a = 54.65$,

$P_a = 165.5$ kN.


Problem 7.11

For the upper bound mechanism of Figure 7.10(a) (a tunnel heading) compute an upper bound value for $q_t$ (internal pressure inside the tunnel applied to the face of the heading). The given data are:
$\alpha = \beta = 60°$, $\delta = 90°$, $H = 10$ m, $D = 2$ m, $c_u = 40$ kPa, $\phi_u = 0$, $\gamma = 20$ kN/m$^3$.

Solution:

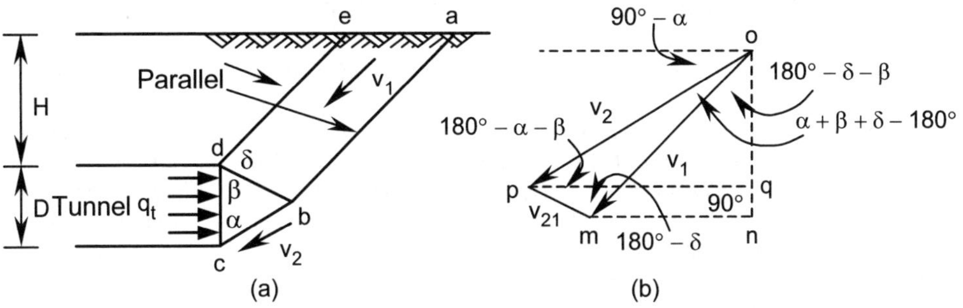

Figure 7.10. Problem 7.11.

From the geometry of the upper bound mechanism and the displacement diagram:

$$bc = bd = cd = 2 \text{ m}, \ de = \frac{10.0}{\cos(180.0° - 60.0° - 90.0°)} = 11.55 \text{ m}.$$

$$ab = \frac{11.0}{\cos(180.0° - 60.0° - 90.0°)} = 12.70 \text{ m}.$$

$$v_1 = 1.0, on = \cos(180.0° - 90.0° - 60.0°) = 0.866,$$

$$nm = \sin(180.0° - 90.0° - 60.0°) = 0.5.$$

$$v_2 = v_1 \frac{\sin(180.0° - \delta)}{\sin(180.0° - \alpha - \beta)} = \frac{1.0 \times \sin 90.0°}{\sin 60.0°} = 1.155.$$

$$oq = 1.155 \cos(60.0° + 60.0° + 90.0° - 180.0°) = 1.155 \times \cos 60.0° = 0.577,$$
$$pq = v_2 \times \sin \alpha = 1.155 \times \sin 60.0° = 1.0.$$

$$v_{21} = v_1 \frac{\sin(\alpha + \beta + \delta - 180.0°)}{\sin(180.0° - \alpha - \beta)} = \frac{1.0 \times \sin 30.0°}{\sin 60.0°} = 0.577.$$

$$w_1 = \frac{2.0(11.55 + 12.70) \times 1.0}{2} \times 20.0 = 485.0 \text{ kN}.$$

$$w_2 = \frac{2.0 \times 2.0 \times \cos 30.0° \times 1.0}{2} \times 20.0 = 34.6 \text{ kN}.$$

Compute external work:

$$E_e = -q_t \times dc \times pq + w_1 \times on + w_2 \times oq,$$
$$E_e = -q_t \times 2.0 \times 1.0 + 485.0 \times 0.866 + 34.6 \times 0.577,$$
$$E_e = 440.0 - 2q_t.$$

Compute internal work:

$$E_i = ed \times c_u \times v_1 + ab \times c_u \times v_1 + bc \times c_u \times v_2 + bd \times c_u \times v_{21},$$
$$E_i = 11.55 \times 40.0 \times 1.0 + 12.70 \times 40.0 \times 1.0 + 2.0 \times 40.0 \times 1.155 + 2.0 \times 40.0 \times 0.577,$$
$$E_i = 1108.6.$$

Equate external work to internal work:

$$440.0 - 2q_t = 1108.6 \rightarrow q_t = -334.3 \text{ kPa (tensile)}.$$

## 7.3 REFERENCES AND RECOMMENDED READINGS

Atkinson, J.H. 1993. *An introduction to the mechanics of soils and foundations*. London: McGraw-Hill.

Aysen, A. 1987. Lower bound solution for soil mechanics problems using finite element method. *Proc. 2nd national conf. SMFE*, 1: 121-136. University of Bogazici, Istanbul, Turkey.

Aysen, A. 2002. *Soil mechanics: Basic concepts and engineering applications*. Lisse: Balkema.

Aysen, A. & Loadwick, F. 1995. Stability of slopes in cohesive frictional soil using upper bound collapse mechanisms and numerical methods. *Proc. 14th Australasian conf. on the mechanics of structures and materials*, 1: 55-59.

Aysen, A. & Sloan, S.W. 1991a. Undrained stability of shallow square tunnel. *Journal GE, ASCE*, 117(8): 1152-1173.

Aysen, A. & Sloan, S. W. 1991b. Stability of a circular tunnel in a cohesive frictional soil. *Proc. 6th intern. conf. in Australia on finite element methods*, 1: 68-76. University of Sydney.

Aysen, A. & Sloan, S.W. 1991c. Undrained stability of a plane strain heading. *Research Report No. 059.02.1991, ISBN 0 7259 07134*. NSW, Australia: The University of Newcastle.

Aysen, A. & Sloan, S.W. 1992. Stability of slopes in cohesive frictional soil. *Proc. 6th Australia-New Zealand conf. on geomechanics*: Geotechnical risk-identification, evaluation and solutions: 414-419. New Zealand: New Zealand Geomechanics Society.

Britto, A.M. & Kusakabe, O. 1985. Upper bound mechanisms for undrained axisymmetric problems. *Proc. 5th intern. conf. on numerical methods in geomechanics*: 1691-1698. Nagoya.

Chen, W.F. 1975. *Limit analysis and soil Plasticity*. Amsterdam: Elsevier.

Chen, W.F. & Baladi, G.Y. 1985. *Soil plasticity: theory and implementation*. Amsterdam: Elsevier.

Davis, E.H., Gunn, M.J., Mair, R.J. & Seneviratne, H.N. 1980. The stability of shallow tunnels and underground openings in cohesive material. *Geotechnique*, 30(4): 397-416.

Heyman, J. 1973. The stability of a vertical cut. *Intern. Journal of mechanical science*, (15): 845-854.

Lysmer, J. 1970. Limit analysis of plane problems in soil mechanics. *Journal GE, ASCE*, 96(4): 1311-1334.

Mulhaus, H.B. 1985. Lower bound solutions for circular tunnels in two and three dimensions. *Journal of rock mech. and rock eng.*, 18: 37-52.

Naylor, D.J. & Pande, G.N. 1981. *Finite elements in geotechnical engineering*. Swansea: Pineridge Press.

Parry, R.H.G. 1995. *Mohr circles, stress paths and geotechnics*. London: E & Spon.

Pastor, J. 1978. Limit analysis: numerical determination of complete statical solutions: application to the vertical cut. *Journal de mecanique appliquee, (in French)*, (2): 167-196.

Sloan, S.W. 1988. Lower bound limit analysis using finite elements and linear programming. *Intern. journal for numerical and analytical methods in geomechanics*, 12(1): 61-77.

Sloan, S.W. 1989. Upper bound limit analysis using finite elements and linear Programming. *Intern. journal for numerical and analytical methods in geomechanics*, 13(3): 263-282.

Sloan, S.W. & Aysen, A. 1992. Stability of shallow tunnels in soft ground. In G.T. Houlsby & A.N. Schofield (eds), *Predictive soil mechanics*. London: Thomas Telford.

Turgeman, S. & Pastor, J. 1982. Limit analysis: a linear formulation of the kinematic approach for axisymmetric mechanic problems. *Intern. journal for numerical and analytical methods in geomechanics*, 6: 109-128.

CHAPTER 8

# Lateral Earth Pressure and Retaining Walls

## 8.1   INTRODUCTION

The problems solved in this chapter are divided into three major categories:
1. The evaluation of the magnitude and distribution of *lateral earth pressure* behind a retaining wall and understanding the concepts of *active* and *passive* failure. Other factors considered are the effects of the surface load and stratified backfill (Problem 8.1), sloping backfill (Problem 8.2), the location of water table (Problem 8.3), and the existence of friction or adhesion on the interface between wall and backfill (Problem 8.4). In the above problems a Rankine type method where the lateral pressure increases linearly with depth is adopted. For this purpose the corresponding active and passive earth pressure coefficients are formulated. For a wall with significant friction, and irregular ground surface, a *Coulomb wedge analysis*, can be applied (Problem 8.5). In this method the actual lateral pressure distribution is not defined but based on experimental data, the location of the resultant of these (lateral) pressures can be obtained. The stability of different types of retaining walls that are subjected to lateral earth pressures are then explained in Problems 8.6 and 8.7.
2. The static analysis of *sheet pile walls* are demonstrated in Problems 8.9 and 8.10 along with determination of the maximum bending moment in the sheet pile and analysis of the anchorage system in an anchored sheet pile.
3. This category (Problem 8.11) investigates the stability of a *reinforced soil* where the lateral earth pressures are resisted by reinforcement elements rather than the retaining wall.

## 8.2   PROBLEMS

Problem 8.1

An 8 m high retaining wall retains a soil comprised of two 4 m thick layers (Figure 8.1(a)) with the following properties:

Upper layer: $c' = 10$ kPa, $\phi' = 18°$, $\gamma = 18$ kN/m$^3$;

lower layer: $c' = 0$, $\phi' = 35°$, $\gamma = 18$ kN/m$^3$.

For a surface load $q = 50$ kPa, determine the active thrust and its distance from the base of the wall.

Solution:

The distribution of lateral active earth pressure behind a smooth vertical wall is:

$$p_a = \sigma'_z k_a - 2c'\sqrt{k_a}$$ (8.1)

where $k_a$ is the lateral active earth pressure coefficient given by:

$$k_a = \frac{1 - \sin\phi'}{1 + \sin\phi'} = \tan^2(45° - \phi'/2)$$ (8.2)

For the passive case:

$$p_p = \sigma'_z k_p + 2c'\sqrt{k_p}$$ (8.3)

where $k_p$ is the lateral passive earth pressure coefficient given by:

$$k_p = \frac{1 + \sin\phi'}{1 - \sin\phi'} = \tan^2(45° + \phi'/2)$$ (8.4)

For layer 1:
$$k_a = \tan^2(45.0° - \phi'/2) = \tan^2(45.0° - 18.0°/2) = 0.528.$$
For layer 2:
$$k_a = \tan^2(45.0° - 35.0°/2) = 0.271.$$
At $z = 0$,

$\sigma'_z = 50.0$ kPa,

$$p_a = 50.0 \times 0.528 - 20.0\sqrt{0.528} = 11.9 \text{ kPa}.$$
At $z = 4.0$ m,
$$\sigma'_z = 18.0 \times 4.0 + 50.0 = 122.0 \text{ kPa},$$

$$p_a = 122.0 \times 0.528 - 20.0\sqrt{0.528} = 49.9 \text{ kPa}.$$
At $z = 4.0$ m but using $k_a$ of the lower layer:

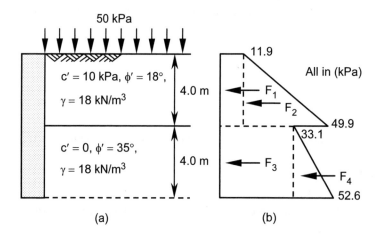

(a)    (b)

Figure 8.1. Problem 8.1.

$p_a = 122.0 \times 0.271 - 0.0\sqrt{0.271} = 33.1$ kPa.

At $z = 8.0$ m,

$\sigma_z' = 18.0 \times 4.0 + 18.0 \times 4.0 + 50.0 = 194.0$ kPa,

$p_a = 194.0 \times 0.271 = 52.6$ kPa.

The lateral earth pressure distribution is shown in Figure 8.1(b) from which:

$F_1 = 11.9 \times 4.0 \times 1.0 = 47.6$ kN,

$F_2 = (49.9 - 11.9) \times 4.0/2 \times 1.0 = 76.0$ kN,

$F_3 = 33.1 \times 4.0 \times 1.0 = 132.4$ kN,

$F_4 = (52.6 - 33.1) \times 4.0/2 \times 1.0 = 39.0$ kN.

$$P_a = \sum_{i=1}^{i=4} F_i,$$

$P_a = 47.6 + 76.0 + 132.4 + 39.0 = 295.0$ kN.

Take the moment of every force about the base to find the position of the resultant:

$$\bar{z} = \frac{47.6(2.0 + 4.0) + 76.0(4.0/3 + 4.0) + 132.4 \times 2.0 + 39.0 \times 4.0/3}{295.0} = 3.42 \text{ m}.$$

Problem 8.2

A retaining wall of 5 m height retains a sloping backfill with $\beta$ (the angle of the sloping ground with horizontal) = 20°. The properties of the backfill are:

$c' = 0$, $\phi' = 35°$, $\gamma = 17$ kN/m$^3$.

Determine the active thrust on the wall and its horizontal and vertical components.

Solution:

For a granular soil with shear strength parameters $c' = 0$, $\phi'$ and $\beta < \phi'$:

$$k_a = \frac{\cos\beta - \sqrt{\cos^2\beta - \cos^2\phi'}}{\cos\beta + \sqrt{\cos^2\beta - \cos^2\phi'}} \qquad (8.5)$$

The active earth pressure at depth $z$ is:

$$p_a = \gamma z k_a \cos\beta \qquad (8.6)$$

The total active thrust is parallel to the ground surface and acts at a height $H/3$ above the base and is given by:

$$P_a = \frac{\gamma H^2}{2} k_a \cos\beta \qquad (8.7)$$

In the passive state the earth pressure coefficient is:

$$k_p = \frac{\cos\beta + \sqrt{\cos^2\beta - \cos^2\phi'}}{\cos\beta - \sqrt{\cos^2\beta - \cos^2\phi'}} \qquad (8.8)$$

The passive earth pressure at depth $z$ and the total passive thrust are calculated using equations similar to Equations 8.6 and 8.7 in which $k_a$ is replaced by $k_p$. In the presence of a surface load $q$, $\gamma z$ must be replaced by $\gamma z + q$.

Using Equation 8.5: $k_a = 0.342$.
The total thrust from Equation 8.7 is:
$$P_a = 0.5 \times 17.0 \times 5.0^2 \times 0.342 \times \cos 20.0° \times 1.0 = 68.3 \text{ kN}.$$

$$P_{ah} = 68.3 \times \cos 20.0° = 64.2 \text{ kN}.$$
$$P_{av} = 68.3 \times \sin 20.0° = 23.4 \text{ kN (downwards)}.$$

Problem 8.3

Re-work Problem 8.1 assuming that the water table is located 2 m below the ground surface (Figure 8.2(a)). The saturated unit weight for both layers is 19.5 kN/m$^3$.

Solution:

At $z = 0$,

$\sigma'_z = 50.0$ kPa,

$$p_a = 50.0 \times 0.528 - 20.0\sqrt{0.528} = 11.9 \text{ kPa}.$$

At $z = 2.0$ m,

$$\sigma'_z = 18.0 \times 2.0 + 50.0 = 86.0 \text{ kPa},$$

$$p_a = 86.0 \times 0.528 - 20.0\sqrt{0.528} = 30.9 \text{ kPa}.$$

At $z = 4.0$ m,

$$\sigma'_z = 18.0 \times 2.0 + 19.5 \times 2.0 + 50.0 - 9.81 \times 2.0 = 105.4 \text{ kPa},$$

$$p_a = 105.4 \times 0.528 - 20.0\sqrt{0.528} = 41.1 \text{ kPa}.$$

Again at $z = 4.0$ m but using $k_a$ of the lower layer:

$$p_a = 105.4 \times 0.271 - 0.0\sqrt{0.271} = 28.6 \text{ kPa}.$$

At $z = 8.0$ m,

$$\sigma'_z = 18.0 \times 2.0 + 19.5 \times 2.0 + 19.5 \times 4.0 - 9.81 \times 6.0 + 50.0 = 144.1 \text{ kPa},$$

$$p_a = 144.1 \times 0.271 = 39.0 \text{ kPa}.$$

The lateral earth pressure distribution is shown in Figure 8.2(b).
Calculate the resultant of active forces:

$$F_1 = 11.9 \times 2.0 \times 1.0 = 23.8 \text{ kN},$$

$$F_2 = (30.9 - 11.9) \times 2.0/2 \times 1.0 = 19.0 \text{ kN},$$

$$F_3 = 30.9 \times 2.0 \times 1.0 = 61.8 \text{ kN},$$

$$F_4 = (41.1 - 30.9) \times 2.0/2 \times 1.0 = 10.2 \text{ kN}.$$

$$F_5 = 28.6 \times 4.0 \times 1.0 = 114.4 \text{ kN},$$

$$F_6 = (39.0 - 28.6) \times 4.0/2 \times 1.0 = 20.8 \text{ kN}.$$

$$P_a = 23.8 + 19.0 + 61.8 + 10.2 + 114.4 + 20.8 = 250.0 \text{ kN}.$$

Take the moment of every force about the base to find the position of the resultant:

$$\bar{z} = \frac{23.8 \times 7.0 + 19.0 \times 6.67 + 61.8 \times 5.0 + 10.2 \times 4.67 + 114.4 \times 2.0 + 20.8 \times 1.33}{250.0} = 3.62 \text{ m}.$$

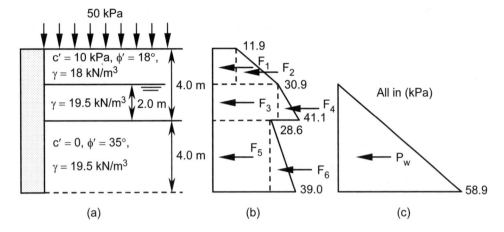

(a)     (b)     (c)

Figure 8.2. Problem 8.3.

Calculate the force due to water pressure behind the wall (Figure 8.2(c)).
$u$ (pore pressure at $z = 8.0$ m) $= 9.81 \times 6.0 = 58.9$ kPa.

$$P_w = \frac{58.9 \times 6.0 \times 1.0}{2} = 176.7 \text{ kN. Thus the total force applied to the retaining wall is:}$$

$P = 250.0 + 176.7 = 426.7$ kN (per meter run of wall).

Problem 8.4

A 10 m retaining wall retains soil with the following properties: $c_u = 40$ kPa, $\gamma = 17.5$ kN/m$^3$, $c_w = 17.6$ kPa, and $\delta'$ (friction angle mobilized between wall and soil) $= 0$. Determine the magnitude of the active thrust:
(a) when the surface carries no surcharge,
(b) when a surface surcharge of 50 kPa is applied.
In both cases the ground surface is horizontal.

Solution:

This problem can be solved by applying the wedge analysis for $c_u$, $\phi_u = 0$ soil which can be formulated for vertical wall with horizontal ground surface.

$$\cot \alpha = \sqrt{1 + \frac{c_w}{c_u}} \tag{8.9}$$

where $\alpha$ is the angle of the failure plane with horizontal:

$$P_a = \frac{\gamma(H^2 - z_o^2)}{2} - 2c_u(H - z_o)\sqrt{1 + \frac{c_w}{c_u}} \qquad \text{for } q = 0 \tag{8.10}$$

If vertical surface loading $q$ exists then:

$$P_a = \frac{\gamma(H^2 - z_o^2)}{2} + q(H - z_o) - 2c_u(H - z_o)\sqrt{1 + \frac{c_w}{c_u}} \qquad \text{for } q > 0 \tag{8.11}$$

The magnitude of $z_o$ is unknown but may be estimated from the following equation which is based on the linear distribution of the lateral active earth pressure behind a smooth wall:

$$z_o = \frac{2c_u}{\gamma} - \frac{q}{\gamma} \tag{8.12}$$

Alternatively we may assume Equation 8.10 to be equivalent to a linear distribution of lateral earth pressure given by:

$$p_a = \gamma z - 2c_u \sqrt{1 + \frac{c_w}{c_u}} \tag{8.13}$$

where $z = 0$ at the ground surface and $z = H$ at the base of the wall. With the assumption of a linear lateral pressure distribution, the depth at which the earth pressure becomes zero is:

$$z_o = \frac{2c_u}{\gamma} \sqrt{1 + \frac{c_w}{c_u}} \tag{8.14}$$

The total active thrust is the integral of Equation 8.13 with a correction for the tension zone (ignoring the tensile area in the stress distribution graph).

$$P_a = \frac{\gamma H(H - z_o)}{2} - c_u(H - z_o)\sqrt{1 + \frac{c_w}{c_u}} \tag{8.15}$$

For $q > 0$, and $p_a < 0$ at $z = 0$, the equivalent linear lateral earth pressure, depth of tension crack and total active thrust are given by:

$$p_a = \gamma z + q - 2c_u \sqrt{1 + \frac{c_w}{c_u}} \tag{8.16}$$

$$z_o = \frac{2c_u}{\gamma} \sqrt{1 + \frac{c_w}{c_u}} - \frac{q}{\gamma} \tag{8.17}$$

$$P_a = \frac{\gamma H(H - z_o)}{2} + \frac{q(H - z_o)}{2} - c_u(H - z_o)\sqrt{1 + \frac{c_w}{c_u}} \tag{8.18}$$

For the passive state:

$$P_p = \frac{\gamma H^2}{2} + qH + 2c_u H \sqrt{1 + \frac{c_w}{c_u}} \tag{8.19}$$

The critical magnitude of $\alpha$ is given by Equation 8.9.
(a) From Equations 8.14 and 8.15 (or 8.10):
$$z_o = (2 \times 40.0/17.5)\sqrt{1 + 17.6/40.0} = 5.48 \text{ m}.$$

$$P_a = 17.5 \times 10.0(10.0 - 5.48)/2 - 40.0(10.0 - 5.48)\sqrt{1 + 17.6/40.0} = 178.5 \text{ kN}.$$

(b) Similarly; and using Equations 8.17 and 8.18 (or 8.11):
$$z_o = (2 \times 40.0/17.5)\sqrt{1 + 17.6/40.0} - 50.0/17.5 = 2.63 \text{ m}.$$
$$P_a = 17.5 \times 10.0(10.0 - 2.63)/2 + 50.0(10.0 - 2.63)/2 -$$
$$40.0(10.0 - 2.63)\sqrt{1 + 17.6/40.0} = 475.4 \text{ kN}.$$

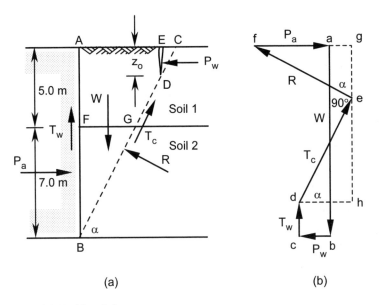

Figure 8.3. Problem 8.5.

Problem 8.5

A retaining wall of height 12 m (Figure 8.3(a)) retains a two-layer soil having the following properties:

0 m - 5 m below the surface: $c_u = 12$ kPa, $\phi_u = 0$, $\gamma = 17$ kN/m$^3$.

Below 5 m, $c_u = 35$ kPa, $\phi_u = 0$, and $\gamma = 18$ kN/m$^3$.

Calculate the magnitude of the total active thrust and the critical value of $\alpha$. For this purpose formulate $P_a$ in terms of the angle $\alpha$ using the force diagram and set:

$$\frac{\partial P_a}{\partial \alpha} = 0. \ \delta' = 0 \text{ and } c_w = 10 \text{ kPa}.$$

Solution:

Depth of tension zone from Equation 8.14:

$$z_o = \frac{2 \times 12.0}{17.0} \sqrt{1 + \frac{10}{12.0}} = 1.91 \text{ m}.$$

Calculate the total weight of the trial wedge using the geometry of the wall and trial edge shown in Figure 8.3(a):

$W = w_1 + w_2 =$ area of $AFGDE \times 1.0 \times \gamma_1 +$ area of $FBG \times 1.0 \times \gamma_2$.

$w_1 =$ (area of $AFGC$ - area of $EDC$) $\times 1.0 \times \gamma_1$,

$$w_1 = \left( \frac{7.0 \cot \alpha + 12.0 \cot \alpha}{2} \times 5.0 - \frac{1.91 \times 1.91 \cot \alpha}{2} \right) \times 17.0 = 776.5 \cot \alpha \text{ kN}.$$

$$w_2 = \frac{7.0 \times 7.0 \cot \alpha}{2} \times 18.0 = 441.0 \cot \alpha \text{ kN}.$$

$W = w_1 + w_2 = 776.5 \cot \alpha + 441.0 \cot \alpha = 1217.5 \cot \alpha.$

Calculate shearing resistances (due to cohesion) on the failure plane and back face of the wall (due to adhesion):

$$T_c = GD \times 1.0 \times c_{u1} + BG \times 1.0 \times c_{u2} = \frac{5.0 - 1.91}{\sin \alpha} \times 12.0 + \frac{7.0}{\sin \alpha} \times 35.0 = \frac{282.1}{\sin \alpha} \text{ kN.}$$

$$T_w = (AB - z_o) \times 1.0 \times c_w = (12.0 - 1.91) \times 10.0 = 100.9 \text{ kN.}$$

Calculate the horizontal force due to water pressure in the tension crack:

$$P_w = \frac{(\gamma_w \times z_o) \times z_o \times 1.0}{2} = \frac{9.81 \times 1.91 \times 1.91}{2} = 17.9 \text{ kN.}$$

From the polygon of the forces shown in Figure 8.3(b) $P_a$ can be evaluated as follows:

$$P_a = fg - ag = ge \tan \alpha - (dh - P_w) = (W - T_w - eh) \tan \alpha - (T_c \cos \alpha - P_w),$$
$$P_a = (W - T_w - T_c \sin \alpha) \tan \alpha - (T_c \cos \alpha - P_w).$$

Substituting values of $W$, $T_w$, $T_c$ and $P_w$:

$$P_a = (1217.5 \cot \alpha - 100.9 - 282.1) \tan \alpha - (282.1 \cot \alpha - 17.9),$$
$$P_a = 1235.4 - 383.0 \tan \alpha - 282.1 \cot \alpha.$$

Calculate the critical value of $\alpha$:

$$\frac{\partial P_a}{\partial \alpha} = -383.0 \times \frac{1}{\cos^2 \alpha} - 282.1 \times \frac{-1}{\sin^2 \alpha} = 0 \rightarrow \frac{\sin^2 \alpha}{\cos^2 \alpha} = \frac{282.1}{383.0},$$

$$\tan \alpha = \frac{282.1}{383.0} = 0.858 \rightarrow \alpha = 40.63°. \text{ Thus}$$

$$P_a = 1235.4 - 383.0 \tan 40.63° - 282.1 \cot 40.63° = 578.0 \text{ kN.}$$

## Problem 8.6

A concrete gravity retaining wall is 6.6 m high and 3.2 m wide. If the thickness of the soil at the front of the wall is 2 m, determine the maximum and minimum base pressures assuming no base friction or adhesion. The soil has the following properties:

$c' = 0$, $\phi' = 35°$, $\rho$ (for soil) = 1.8 Mg/m$^3$. $\rho$ (for concrete) = 2.4 Mg/m$^3$.

Solution:

Calculate $k_a$ and $k_p$ from Equations 8.2 and 8.4:

$$k_a = \tan^2(45.0° - 35.0°/2) = 0.271.$$
$$k_p = \tan^2(45.0° + 35.0°/2) = 3.690.$$

Active pressure distribution and thrust:

At $z = 0$, $\sigma'_z = 0.0$ and $p_a = 0.0 \times 0.271 - 0.0\sqrt{0.271} = 0.0$.

At $z = 6.6$ m, $\sigma'_z = 1.8 \times 9.81 \times 6.6 = 116.5$ kPa,

$$p_a = 116.5 \times 0.271 - 0.0\sqrt{0.271} = 31.6 \text{ kPa.}$$

$$P_a = \frac{31.6 \times 6.6 \times 1.0}{2} = 104.3 \text{ kN.}$$

Passive pressure distribution and thrust:

At $z = 0$, $\sigma'_z = 0.0$ and $p_p = 0.0 \times 3.690 + 0.0\sqrt{3.690} = 0.0$.

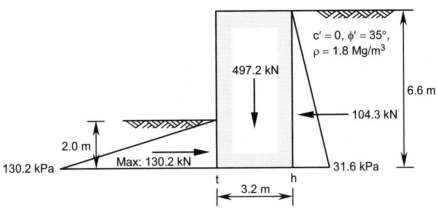

Figure 8.4. Problem 8.6.

At $z = 2.0$ m, $\sigma_z' = 1.8 \times 9.81 \times 2.0 = 35.3$ kPa,

$p_p = 35.3 \times 3.690 + 0.0\sqrt{3.690} = 130.2$ kPa.
$P_p = 130.2 \times 2.0 \times 1.0 / 2 = 130.2$ kN.

Factor of safety against sliding is the ratio of the resisting forces in the sliding direction to the component of the active thrust or disturbing forces along the same direction:

$F_S = 130.2 / 104.3 = 1.248 \approx 1.25$. The weight of the wall is:

$W = 6.6 \times 3.2 \times 1.0 \times 2.4 \times 9.81 = 497.2$ kN.

Taking the moments of the forces about the toe (point $t$, Figure 8.4) we have:
$(130.2 / F_S) \times (2.0/3) + 497.2 \times 1.6 - 104.3 \times (6.6/3) - N \times x = 0$.

Where $N$ is the total vertical force (or vertical reaction of the soil applied to the base) and $x$ is the distance of this force from the toe.

From vertical equilibrium $N = 497.2$ kN; thus
$(130.2 / 1.248) \times (2.0/3) + 497.2 \times 1.6 - 104.3 \times (6.6/3) - 497.2 \times x = 0$,

$x = 1.278$ m.

The eccentricity $e$ is calculated from:

$$e = \frac{L}{2} - x \qquad (8.20)$$

Thus $e = 1.6 - 1.278 = 0.322$ m.

For a rigid rectangular footing the contact pressure distribution under the footing may be assumed linear. With the symmetric loading about the $L$-axis, the contact pressures at the two edges of the rectangle are (Equations 5.51):

$$q_{max} = \frac{N}{LB}\left(1 + \frac{6e}{L}\right), \quad q_{min} = \frac{N}{LB}\left(1 - \frac{6e}{L}\right) \qquad (8.21)$$

For $e > 0$ the maximum contact pressure occurs at the toe whilst the minimum contact pressure occurs at the heel:

$q_{max} = (497.2 / 3.2 \times 1.0)(1.0 + 6 \times 0.322 / 3.2) = 249.2$ kPa (at point $t$, Figure 8.4).
$q_{min} = (497.2 / 3.2 \times 1.0)(1.0 - 6 \times 0.322 / 3.2) = 61.6$ kPa (at point $h$, Figure 8.4).

Problem 8.7

For a concrete gravity retaining wall of 3.2 m width and 6.6 m height, determine the maximum and minimum base pressures assuming a base friction angle of $\delta' = 15°$, and a base adhesion of $c_b = 10$ kPa. A surcharge load of 20 kPa is applied vertically to the ground surface on the backfill. The soil has the following properties:
$c' = 10$ kPa, $\phi' = 25°$, $\rho$ (for soil) $= 1.8$ Mg/m$^3$, $\rho$ (for concrete) $= 2.4$ Mg/m$^3$.

Solution:

Calculate $k_a$ and $k_p$ from Equations 8.2 and 8.4:

$k_a = \tan^2(45.0° - 25.0°/2) = 0.406$, $k_p = \tan^2(45.0° + 25.0°/2) = 2.464$.

Active pressure distribution and thrust:

At $z = 0$, $\sigma'_z = 20.0$ and $p_a = 20.0 \times 0.406 - 2 \times 10.0\sqrt{0.406} = -4.6$ kPa.

The depth of tension crack is calculated by setting Equation 8.1 to zero (or use Equation 8.17 with $c_w = 0$, and replace $c_u$ with $c'$):

$p_a = \sigma'_z k_a - 2c'\sqrt{k_a} = (\gamma z_o + q)k_a - 2c'\sqrt{k_a} = 0$, thus

$$z_o = \frac{2c'}{\gamma\sqrt{k_a}} - \frac{q}{\gamma} = \frac{2c'\tan(45° + \phi'/2)}{\gamma} - \frac{q}{\gamma} \quad (8.22)$$

$z_o = 2 \times 10.0/(1.8 \times 9.81\sqrt{0.406}) - 20.0/(1.8 \times 9.81) = 0.64$ m.

At $z = 6.6$ m, $\sigma'_z = 1.8 \times 9.81 \times 6.6 + 20.0 = 136.5$ kPa,

$p_a = 136.5 \times 0.406 - 2 \times 10.0\sqrt{0.406} = 42.7$ kPa.

$P_a = \dfrac{42.7 \times (6.6 - 0.64) \times 1.0}{2} = 127.2$ kN.

Passive pressure distribution and thrust:

At $z = 0$, $\sigma'_z = 0.0$ and $p_p = 0.0 \times 2.464 + 2 \times 10.0\sqrt{2.464} = 31.4$ kPa.

At $z = 2.0$ m, $\sigma'_z = 1.8 \times 9.81 \times 2.0 = 35.3$ kPa,

$p_p = 35.3 \times 2.464 + 2 \times 10.0\sqrt{2.464} = 118.4$ kPa.

$P_{p1} = 31.4 \times 2.0 \times 1.0 = 62.8$ kN.

$P_{p2} = (118.4 - 31.4) \times 2.0 \times 1.0/2 = 87.0$ kN.

$P_p = 62.8 + 87.0 = 149.8$ kN.

The active and passive pressure diagrams are shown in Figure 8.5. The weight of the wall is 497.2 kN. Factor of safety against sliding is:

$$F_S = \frac{P_p + W \times \tan\delta' + L \times 1.0 \times c_b}{P_a},$$

$F_S = (149.8 + 497.2 \times \tan 15.0° + 3.2 \times 1.0 \times 10.0)/127.2 = 2.477$.

Taking the moments of the forces about the toe:

$(62.8/2.477) \times (2.0/2) + (87.0/2.477) \times (2.0/3) + 497.2 \times 1.6 -$

$127.2 \times (6.6 - 0.64)/3 - 497.2 \times x = 0 \rightarrow x = 1.190$ m.

The eccentricity $e$ is calculated as:

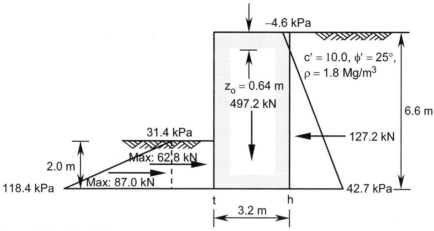

Figure 8.5. Problem 8.7.

$e = 1.6 - 1.19 = 0.41\,\text{m}$.

$$q_{max} = \frac{497.2}{3.2 \times 1.0}\left(1.0 + \frac{6 \times 0.41}{3.2}\right) = 274.8\,\text{kPa (at point } t\text{, Figure 8.5)}.$$

$$q_{min} = \frac{497.2}{3.2 \times 1.0}\left(1.0 - \frac{6 \times 0.41}{3.2}\right) = 35.9\,\text{kPa (at point } h\text{, Figure 8.5)}.$$

## Problem 8.8

The concrete gravity retaining wall shown in Figure 8.6 supports two layers of soil each having a thickness of 3 m. The properties of the layers are:

upper layer: $c' = 0$, $\phi' = 30°$, $\gamma_{dry} = 17.5\,\text{kN/m}^3$, and $\gamma_{sat} = 19.5\,\text{kN/m}^3$;

lower layer: $c' = 10\,\text{kPa}$, $\phi' = 18°$, and $\gamma_{sat} = 19\,\text{kN/m}^3$. There is a surface load of 50 kPa and the water table is 1.5 m below the ground surface. The front of the wall is supported by soil with $c' = 20\,\text{kPa}$, $\phi' = 25°$ and $\gamma = 18\,\text{kN/m}^3$. Determine:

(a) the factor of safety against sliding assuming that the cohesion between the base of the wall and the soil is 20 kPa, and the mobilized friction angle on this interface is 25°,

(b) the factor of safety against overturning,

(c) the distribution of the contact pressure under the base of the wall.

Take the unit weight of the concrete as 24 kN/m³. Assume the back and front faces of the wall are smooth ($c_w = 0$, $\delta'$ (wall) $= 0$).

Solution:

(a) For the upper layer (Equation 8.2): $k_a = \tan^2(45.0° - 30.0°/2) = 0.333$.

For the lower layer: $k_a = \tan^2(45.0° - 18.0°/2) = 0.528$.

At $z = 0$, $\sigma_z = 50.0\,\text{kPa}$, $u = 0$, $\sigma'_z = 50.0\,\text{kPa}$, $p_a = \sigma'_z k_a = 50.0 \times 0.333 = 16.7\,\text{kPa}$.

At $z = 1.5\,\text{m}$, kPa, $u = 0$, $\sigma_z = 50.0 + 17.5 \times 1.5 = 76.2\,\text{kPa}$, $\sigma'_z = 76.2\,\text{kPa}$,

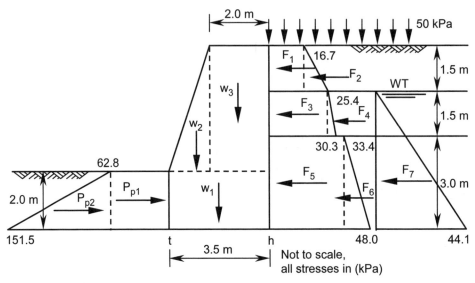

Figure 8.6. Problem 8.8.

$p_a = 76.2 \times 0.333 = 25.4$ kPa.

At $z = 3.0$ m, $\sigma_z = 50.0 + 17.5 \times 1.5 + 19.5 \times 1.5 = 105.5$ kPa, $u = 9.81 \times 1.5 = 14.7$ kPa,

$\sigma'_z = 105.5 - 14.7 = 90.8$ kPa, using $k_a = 0.333$, $p_a = \sigma'_z k_a = 90.8 \times 0.333 = 30.3$ kPa.

Using $k_a = 0.528$, $p_a = \sigma'_z k_a - 2c' \times \sqrt{k_a} = 90.8 \times 0.528 - 2 \times 10.0\sqrt{0.528} = 33.4$ kPa.

At $z = 6.0$ m, $\sigma_z = 50.0 + 17.5 \times 1.5 + 19.5 \times 1.5 + 19.0 \times 3.0 = 162.5$ kPa,

$u = 9.81 \times 4.5 = 44.1$ kPa, $\sigma'_z = 162.5 - 44.1 = 118.4$ kPa,

$p_a = 118.4 \times 0.528 - 2 \times 10.0\sqrt{0.528} = 48.0$ kPa.

The results are shown in Figure 8.6 and the computations are summarized in the table below. Distance of the total horizontal thrust from the base $= 640.85 / 294.66 = 2.175$ m.

| Force (kN) | Arm above the base (m) | Force × arm (kN.m) |
|---|---|---|
| $F_1 = 16.7 \times 1.5 \times 1.0$ (per meter run) $= 25.05$ | 5.25 | 131.51 |
| $F_2 = (25.4 - 16.7) \times 1.5 \times 1/2 \times 1.0 = 6.52$ | 5.00 | 32.60 |
| $F_3 = 25.4 \times 1.5 \times 1.0 = 38.10$ | 3.75 | 142.87 |
| $F_4 = (30.3 - 25.4) \times 1.5 \times 1/2 \times 1.0 = 3.67$ | 3.50 | 12.84 |
| $F_5 = 33.4 \times 3.0 \times 1.0 = 100.20$ | 1.50 | 150.30 |
| $F_6 = (48.0 - 33.4) \times 3 \times 1/2 \times 1.0 = 21.90$ | 1.00 | 21.90 |
| $F_7 = 44.1 \times 4.5 \times 1/2 \times 1.0 = 99.22$ | 1.50 | 148.83 |
| Total thrust $= 294.66 \approx 294.7$ | | Total $= 640.85$ |

Passive pressure distribution and thrust:

$k_p = \tan^2(45.0° + 25.0°/2) = 2.464$.

At $z = 0$, $\sigma'_z = 0.0$ and $p_p = 0.0 \times 2.464 + 2 \times 20.0\sqrt{2.464} = 62.8$ kPa.

At $z = 2.0$ m, $\sigma'_z = 18.0 \times 2.0 = 36.0$ kPa,

$P_p = 36.0 \times 2.464 + 2 \times 20.0\sqrt{2.464} = 151.5$ kPa.

$P_{p1} = 62.8 \times 2.0 \times 1.0 = 125.6$ kN.

$P_{p2} = (151.5 - 62.8) \times 2.0 \times 1.0/2 = 88.7$ kN, $P_p = 125.6 + 88.7 = 214.3$ kN.

Referring to Figure 8.6 the weight of the wall is: $W = w_1 + w_2 + w_3$,

$W = 3.5 \times 2.0 \times 1.0 \times 24.0 + 1.5 \times 4.0 \times 1.0/2 \times 24.0 + 2.0 \times 4.0 \times 1.0 \times 24.0$,

$W = w_1 + w_2 + w_3 = 168.0 + 72.0 + 192.0 = 432.0$ kN.

Factor of safety against sliding is:

$$F_S = \frac{P_p + W \times \tan \delta' + L \times 1.0 \times c_b}{P_a + P_w},$$

$$F_S = \frac{214.3 + 432.0 \times \tan 25.0° + 3.5 \times 1.0 \times 20.0}{294.7} = 1.648 \approx 1.65.$$

(b) For overturning, the factor of safety is defined as the ratio of the sum of resisting moments to the sum of disturbing moments about the toe of the retaining wall.

$$F_V = \frac{\sum M_r}{\sum M_d} \tag{8.23}$$

$$\sum M_r = \frac{125.6 \times 2.0}{2} + \frac{88.7 \times 2.0}{3} + 168.0 \times 1.75 + 72.0 \times 1.0 + 192.0 \times 2.5 = 1030.73 \text{ kN.m.}$$

$\sum M_d = 640.85$ kN.m (see the table on the previous page).

$$F_V = \frac{\sum M_r}{\sum M_d} = \frac{1030.73}{640.85} = 1.608 \approx 1.61.$$

(c) Taking the moments of the forces about the toe:

$(125.6/1.648) \times (2.0/2) + (88.7/1.648) \times (2.0/3) + 168.0 \times 1.75 + 72.0 \times 1.0 +$
$192.0 \times 2.5 - 640.85 - 432.0 \times x = 0 \rightarrow x = 0.734$ m.

$e = 1.75 - 0.734 = 1.016$ m.

$$q_{max} = \frac{432.0}{3.5 \times 1.0}\left(1.0 + \frac{6 \times 1.016}{3.5}\right) = 338.4 \text{ kPa (at point } t, \text{ Figure 8.6).}$$

$$q_{min} = \frac{432.0}{3.5 \times 1.0}\left(1.0 - \frac{6 \times 1.016}{3.5}\right) = -91.5 \text{ kPa (at point } h, \text{ Figure 8.6).}$$

## Problem 8.9

A cantilever sheet pile supports a 6 m high backfill with the following properties:

0 m - 2 m: $c' = 0$, $\phi' = 30°$, $\gamma = 16.5$ kN/m$^3$, 2 m - 4 m: $c' = 0$, $\phi' = 35°$, $\gamma = 17$ kN/m$^3$,
4 m - 6 m: $c' = 15$ kPa, $\phi' = 20°$, $\gamma = 17$ kN/m$^3$.

Embedment depth $D = 3.5$ m and the soil under the dredge line (at both sides) is the same as the soil under the 4 m depth. Determine the factor of safety (in terms of $c'$ and $k_p$) assuming a simplified pressure diagram. The case of the cantilever sheet pile in $c'$, $\phi'$ soil is not formulated in this chapter.

Solution:

In $c' = 0$, $\phi'$ soil values of the earth pressure coefficients $k_a$ and $k_p$ are calculated from the following two equations that are obtained using wedge analysis.

$$k_a = \frac{\sin^2(\theta - \phi')}{\sin^2\theta \sin(\theta + \delta')\left[1 + \sqrt{\dfrac{\sin(\phi' + \delta')\sin(\phi' - \beta)}{\sin(\theta + \delta')\sin(\theta - \beta)}}\right]^2} \tag{8.24}$$

where $\theta$ is the angle of the back face of the wall from the horizontal, $\beta$ is the inclination of the upper ground surface from the horizontal and $\delta'$ is the friction angle mobilized on the soil wall interface. For a vertical wall ($\theta = 90°$) and a horizontal ground surface ($\beta = 0$):

$$k_a = \left[\frac{\cos\phi'}{\sqrt{\cos\delta'} + \sqrt{\sin(\phi' + \delta')\sin\phi'}}\right]^2 \tag{8.25}$$

For a smooth ($\delta' = 0$) vertical wall with horizontal ground surface, $k_a$ becomes identical to Equation 8.2.

$$k_p = \frac{\sin^2(\theta + \phi')}{\sin^2\theta \sin(\theta - \delta')\left[1 - \sqrt{\dfrac{\sin(\phi' + \delta')\sin(\phi' + \beta)}{\sin(\theta - \delta')\sin(\theta - \beta)}}\right]^2} \tag{8.26}$$

For a vertical wall ($\theta = 90°$) and a horizontal ground surface ($\beta = 0$):

$$k_p = \left[\frac{\cos\phi'}{\sqrt{\cos\delta'} - \sqrt{\sin(\phi' + \delta')\sin\phi'}}\right]^2 \tag{8.27}$$

For a smooth ($\delta' = 0$) vertical wall with a horizontal ground surface, this equation reduces to Equation 8.4. In $c'$, $\phi'$ soil the concept of a linear earth pressure distribution can be employed in the following form:

$$p_a = (\gamma z + q)k_a - 2c'\sqrt{k_a\left(1 + \frac{c_w}{c'}\right)} \tag{8.28}$$

where $k_a$ is defined by Equation 8.24 and $c_w$ is the cohesion mobilized on the soil wall interface. For the passive case the linear pressure distribution is given by:

$$p_p = (\gamma z + q)k_p + 2c'\sqrt{k_p\left(1 + \frac{c_w}{c'}\right)} \tag{8.29}$$

Note that the angle of the active and passive thrusts with horizontal is $\delta'$. In this example we assume a value of $\delta' = 20°$ in both active and passive sides with $c_w = 0$. In a simplified pressure distribution behind a cantilever sheet pile it is assumed the active state is created in the backfill along the entire length of the sheet pile with a single force representing the passive state (due to the rotation of the sheet pile) acting at the end point of the sheet pile (Figure 8.7). The front of the sheet pile is in the passive state, which is mobilized to maintain static equilibrium.

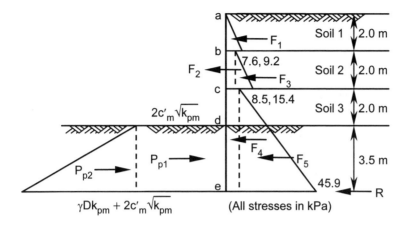

Figure 8.7. Problem 8.9.

Using Equation 8.25 the active earth pressure coefficients are calculated as follows.
For layer 1: $k_a = 0.297$; layer 2: $k_a = 0.245$ and for layer 3: $k_a = 0.427$.
At $z = 0$, $\sigma'_z = 0$, $p_a = \sigma'_z k_a = 0$.
At $z = 2.0$ m, $\sigma'_z = 16.5 \times 2.0 = 33.0$ kPa,
$p_a = 33.0 \times 0.297 = 9.8 \rightarrow p_{ah} = 9.8 \times \cos 20.0° = 9.2$ kPa.
$F_1 = 9.2 \times 2.0/2 \times 1.0 = 9.2$ kN, arm above the base is 8.167 m.
At $z = 2.0$ m, using $k_a = 0.245$,
$p_a = 33.0 \times 0.245 = 8.1 \rightarrow p_{ah} = 8.1 \times \cos 20.0° = 7.6$ kPa.
At $z = 4.0$ m, $\sigma'_z = 16.5 \times 2.0 + 17.0 \times 2.0 = 67.0$ kPa,
$p_a = 67.0 \times 0.245 = 16.4 \rightarrow p_{ah} = 16.4 \times \cos 20.0° = 15.4$ kPa.
$F_2 = 7.6 \times 2.0 \times 1.0 = 15.2$ kN, arm above the base is 6.5 m.
$F_3 = (15.4 - 7.6) \times 2.0/2 \times 1.0 = 7.8$ kN, arm above the base is 6.167 m.
At $z = 4.0$ m, using $k_a = 0.427$,
$p_a = 67.0 \times 0.427 - 2.0 \times 15.0\sqrt{0.427} = 9.0 \rightarrow p_{ah} = 9.0 \times \cos 20.0° = 8.5$ kPa.
At $z = 9.5$ m, $\sigma'_z = 16.5 \times 2.0 + 17.0 \times 2.0 + 17.0 \times 5.5 = 160.5$ kPa,
$p_a = 160.5 \times 0.427 - 2 \times 15.0\sqrt{0.427} = 48.9 \rightarrow p_{ah} = 48.9 \times \cos 20.0° = 45.9$ kPa.
$F_4 = 8.5 \times 5.5 \times 1.0 = 46.7$ kN, arm above the base is 2.75 m.
$F_5 = (45.9 - 8.5) \times 5.5/2 \times 1.0 = 102.8$ kN, arm above the base is 1.833 m.
The results are shown in Figure 8.7.
Passive pressure distribution and thrust:
The value of $k_p$ is calculated from Equation 8.27: $k_p = 3.525$. The mobilized cohesion and
earth pressure coefficient are shown by $c'_m$ and $k_{pm}$ respectively:
At $z = 0$, $\sigma'_z = 0.0$ and
$$p_p = 2 \times c'_m \sqrt{k_{pm}} \rightarrow p_{ph} = 2 \times c'_m \sqrt{k_{pm}} \times \cos 20.0° = 1.88 c'_m \sqrt{k_{pm}} \text{ kPa.}$$
At $z = 3.5$ m, $\sigma'_z = 17.0 \times 3.5 = 59.5$ kPa,

$$P_p = 59.5k_{pm} + 2 \times c'_m \sqrt{k_{pm}} \text{ kPa.}$$

$$P_{ph} = (59.5k_{pm} + 2 \times c'_m \sqrt{k_{pm}}) \cos 20.0° = 55.9k_{pm} + 1.88c'_m \sqrt{k_{pm}} \text{ kPa.}$$

$$P_{p1} = 1.88 \times c'_m \sqrt{k_{pm}} \times 3.5 \times 1.0 = 6.58c'_m \sqrt{k_{pm}} \text{ kN.}$$

$$P_{p1} = 6.58 \times 15.0 / F \sqrt{3.525 / F} = 185.3\sqrt{1/F^3} \text{ kN.}$$

$$P_{p2} = 55.9k_{pm} \times 3.5 / 2 \times 1.0 = 97.8k_{pm} \text{ kN.}$$

$$P_{p2} = 97.8 \times 3.525 / F = 344.7 / F \text{ kN.}$$

Taking the moments of the forces about point $e$:

$$9.2 \times 8.167 + 15.2 \times 6.5 + 7.8 \times 6.167 + 46.7 \times 2.75 + 102.8 \times 1.833 -$$

$$185.3\sqrt{1/F^3} \times 1.75 - 344.7 / F \times 1.167 = 0,$$

$$538.9 - 324.3\sqrt{1/F^3} - 402.3 / F = 0. \text{ A trial and error method gives } F \approx 1.28.$$

## Problem 8.10

A cantilever sheet pile supports a 9 m high backfill with the following properties:
0 m - 3 m: $c' = 0$, $\phi' = 35°$, $\gamma = 18$ kN/m$^3$,
3 m - 6 m: $c' = 15$ kPa, $\phi' = 20°$, $\gamma_{sat} = 20.3$ kN/m$^3$,
6 m - 9 m: $c' = 0$, $\phi' = 35°$, $\gamma_{sat} = 21.1$ kN/m$^3$.
There is a vertical surface load of 20 kPa applied at the backfill. The water table is 3 m below the ground surface of the backfill and has the same level at the front of the sheet pile. The soil under the dredge line (at both sides) is purely cohesive soil with: $c_u = 100$ kPa, and $\phi_u = 0$.
(a) Determine the embedment depth $D$ assuming that the full passive resistance is mobilized,
(b) add a horizontal anchor rod at a depth of 1.5 m and, with the same embedment depth of part (a), calculate the factor of safety (in terms of mobilized cohesion) and the anchor rod force,
(c) determine the location and magnitude of the maximum bending moment,
(d) if the anchor rod is supported by a concrete block anchor with thickness of 0.5 m, width of 2 m (parallel to the sheet pile), and height of 1.5 m, calculate the distance between anchor rods along the sheet pile (anchor rod is anchored at the centre point of the concrete block).
Assume no cohesion and friction resistance along the surfaces of the anchor. Include the surface load in both the active and passive thrusts and assume a factor of safety of 1.5 for the mobilized $k_p$ at the front of the anchor.

Solution:

(a) Using Equation 8.2: for layer 1: $k_a = 0.271$; for layer 2: $k_a = 0.490$ and for layer 3: $k_a = 0.271$.
At $z = 0$, $\sigma'_z = 20.0$ kPa,
$$p_a = 20.0 \times 0.271 = 5.4 \text{ kPa.}$$
At $z = 3.0$ m, $\sigma'_z = 20.0 + 18.0 \times 3.0 = 74.0$ kPa,
$$p_a = 74.0 \times 0.271 = 20.0 \text{ kPa.}$$

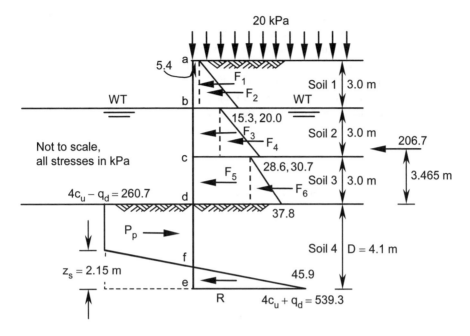

Figure 8.8. Problem 8.10: part (a).

$F_1 = 5.4 \times 3.0 \times 1.0 = 16.2$ kN, arm above the dredge line is 7.5 m.

$F_2 = (20.0 - 5.4) \times 3.0 / 2 \times 1.0 = 21.9$ kN, arm above the dredge line is 7.0 m.

At $z = 3.0$ m, using $k_a = 0.490$,

$p_a = 74.0 \times 0.490 - 2 \times 15.0 \sqrt{0.490} = 15.3$ kPa.

At $z = 6.0$ m, $\sigma'_z = 74.0 + 20.3 \times 3.0 - 9.81 \times 3.0 = 105.5$ kPa,

$p_a = 105.5 \times 0.490 - 2 \times 15.0 \sqrt{0.490} = 30.7$ kPa.

$F_3 = 15.3 \times 3.0 \times 1.0 = 45.9$ kN, arm above the dredge line is 4.5 m.

$F_4 = (30.7 - 15.3) \times 3.0 / 2 \times 1.0 = 23.1$ kN, arm above the dredge line is 4.0 m.

At $z = 6.0$ m, using $k_a = 0.271$,

$p_a = 105.5 \times 0.271 = 28.6$ kPa.

At $z = 9.0$ m, $\sigma'_z = 74.0 + 20.3 \times 3.0 + 21.1 \times 3.0 - 9.81 \times 6.0 = 139.3$ kPa.

$p_a = 139.3 \times 0.271 = 37.8$ kPa.

$F_5 = 28.6 \times 3.0 = 85.8$ kN, arm above the dredge line is 1.5m.

$F_6 = (37.8 - 28.6) \times 3.0 / 2 \times 1.0 = 13.8$ kN, arm above the dredge line is 1.0 m.

The results are shown in Figure 8.8.

Calculation of resultant $R_a$ and its position above the dredge line:

$R_a = F_1 + F_2 + F_3 + F_4 + F_5 + F_6 = 16.2 + 21.9 + 45.9 + 23.1 + 85.8 + 13.8 = 206.7$ kN.

Taking the moments of the forces about the dredge line:

$16.2 \times 7.5 + 21.9 \times 7.0 + 45.9 \times 4.5 + 23.1 \times 4.0 + 85.8 \times 1.5 + 13.8 \times 1.0 = 206.7\bar{z}$,

$\bar{z} = 3.465$.

When the material below the dredge line is a purely cohesive soil with $c_u$ and $\phi_u = 0$, the net pressure diagram under the dredge line will be as shown in Figure 8.8. Noting that in the clay layer $k_a = k_p = 1$, the net pressure at $d$ becomes $4c_u - q_d$, where $q_d$ is the effective vertical stress at the dredge line level at the back of the sheet pile. As the active and passive pressure coefficients under the dredge line are equal, the effect of the weight disappears and the net pressure below point $d$ remains constant up to the point of rotation (of the sheet pile). It can be shown that the net pressure at the end point at the back of the wall is $4c_u + q_d$. For equilibrium of the horizontal forces and moments about the base:

$$z_s = \frac{D(4c_u - q_d) - R_a}{4c_u} \tag{8.30}$$

$$D^2 - \frac{2R_a}{4c_u - q_d}D - \frac{R_a(12c_u\bar{z} + R_a)}{(4c_u - q_d)(2c_u + q_d)} = 0 \tag{8.31}$$

For the simplified pressure diagram with $z_s = 0$, and with $R$ acting as a concentrated force at point $e$:

$$D^2 - \frac{2R_a}{4c_u - q_d}D - \frac{2R_a\bar{z}}{4c_u - q_d} = 0 \tag{8.32}$$

The computed $D$ values are increased by 20% to 40% or, alternatively, the mobilized cohesion $c_{um} = c_u / F$ is substituted for $c_u$ where $F = 1.5$ to 2.0. For a stable sheet pile $4c_{um} - q_d > 0$ or:

$$4c_u / F > q_d \tag{8.33}$$

Calculate the necessary coefficients:

$q_d = \sigma'_z$ at $z = 9$ m $= 139.3$ kPa.

$4c_u - q_d = 4 \times 100.0 - 139.3 = 260.7$ kPa.

$$\frac{2R_a}{4c_u - q_d} = \frac{2 \times 206.7}{260.7} = 1.586, \quad 4c_u + q_d = 539.3 \text{ kPa.}$$

$$\frac{R_a(12c_u\bar{z} + R_a)}{(4c_u - q_d)(2c_u + q_d)} = \frac{206.7(12 \times 100.0 \times 3.465 + 206.7)}{260.7(2 \times 100.0 + 139.3)} = 10.208.$$

Substituting into Equation 8.31:

$D^2 - 1.586D - 10.208 = 0$,

$D = 4.08$ m $\approx 4.10$ m.

Using Equation 8.30:

$z_s = (4.10 \times 260.7 - 206.7)/(4 \times 100.0) = 2.15$ m.

From the similarity concept:

$$\frac{ef}{4c_u + q_d} = \frac{z_s}{(4c_u + q_d) + (4c_u - q_d)}, \quad \frac{ef}{539.3} = \frac{2.15}{800.0} \rightarrow ef = 1.45 \text{ m.}$$

$$P_p = 260.7 \times \frac{(4.10 - 2.15) + (4.10 - 1.45)}{2} \times 1.0 = 599.6 \text{ kN.}$$

$R = (539.3 \times 1.45)/2 \times 1.0 = 391.0$ kN. Check statics:

$\sum F(\text{horizontal}) = 599.6 - 206.7 - 391.0 = 1.9$ kN,

which is due to the rounding off of the embedment depth and $z_s$.

(b) With the net pressure diagram shown in Figure 8.9 and full mobilization of $c_u$, the sum of the moments about the anchor rod level yields:

$$D^2 + 2H_{ar}D - \frac{2R_a z'}{4c_u - q_d} = 0 \qquad (8.34)$$

Equilibrium of the horizontal forces is used to determine the force in the anchor rod. The general expression for the mobilized passive thrust is:

$$P_p = D\left(\frac{4c_u}{F} - q_d\right) \qquad (8.35)$$

$z' = 9.0 - 3.465 - 1.5 = 4.035$ m,

$H_{ar} = 9.0 - 1.5 = 7.5$ m.

Use Equation 8.34:

$$4.1^2 + 2 \times 7.5 \times 4.1 - \frac{2 \times 206.7 \times 4.035}{4 \times 100/F - 139.3} = 0,$$

$F = 2.49.$

From Equation 8.35:

$$P_p = 4.10\left(\frac{4 \times 100.0}{2.49} - 139.3\right) = 87.5 \text{ kN},$$

$$R_{ar} = R_a - P_p = 206.7 - 87.5 = 119.2 \text{ kN}.$$

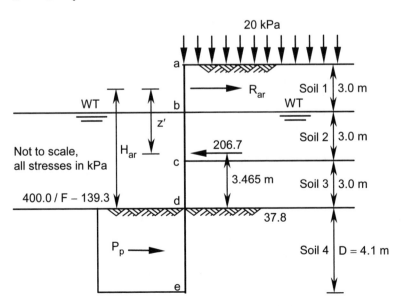

Figure 8.9. Problem 8.10: part (b).

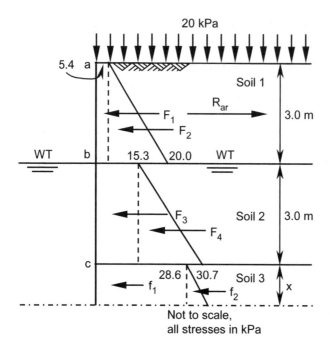

20 kPa

5.4   a

Soil 1

$R_{ar}$

$F_1$                              3.0 m

$F_2$

WT     b        15.3     20.0     WT

$F_3$              Soil 2      3.0 m

$F_4$

c

28.6    30.7     Soil 3

$f_1$                    X

$f_2$

Not to scale,
all stresses in kPa

Figure 8.10. Problem 8.10: part (c).

(c) The maximum bending moment occurs at a point where the shear force is zero.
The sum of $F_1$ to $F_4$ is less than the anchor rod force:

$F_1 + F_2 + F_3 + F_4 = 16.2 + 21.9 + 45.9 + 23.1 = 107.1 \text{ kN} < 119.2 \text{ kN}.$

Thus the position of the zero shear force is located below point c with a depth x measured
from the top of the soil 3 (Figure 8.10). The lateral earth pressure at this depth is:

$p_a = 28.6 + (\gamma - \gamma_w)xk_a,$

$p_a = 28.6 + (21.1 - 9.81)\overset{..}{x} \times 0.271 = 28.6 + 3.06x \text{ kPa}.$

The corresponding forces $f_1$ and $f_2$ are:

$f_1 = 28.6 \times x \times 1.0 = 28.6x \text{ kN},$

$f_2 = 3.06x \times x / 2 \times 1.0 = 1.53x^2 \text{ kN}.$

Set the shear force to zero:

$SF = -R_{ar} + F_1 + F_2 + F_3 + F_4 + f_1 + f_2 = 0, \ -119.2 + 107.1 + 28.6x + 1.53x^2 = 0,$

$1.53x^2 + 28.6x - 12.1 = 0 \rightarrow x = 0.41 \text{ m}.$

$f_1 = 28.6 \times 0.41 = 11.7 \text{ kN},$

$f_2 = 1.53 \times 0.41^2 = 0.3 \text{ kN}.$

The magnitude of the bending moment at this point is:

$BM = R_{ar} \times 4.91 - F_1 \times 4.91 - F_2 \times 4.41 - F_3 \times 1.91 - F_4 \times 1.41 - f_1 \times x / 2 - f_2 \times x / 3,$

$BM = 119.2 \times 4.91 - 16.2 \times 4.91 - 21.9 \times 4.41 - 45.9 \times 1.91 - 23.1 \times 1.41 -$
$11.7 \times 0.41 / 2 - 0.3 \times 0.41 / 3 = 286.5 \text{ kN.m}.$

This means that the back of the sheet pile at this point will be in compression, while its
front will be in tension.

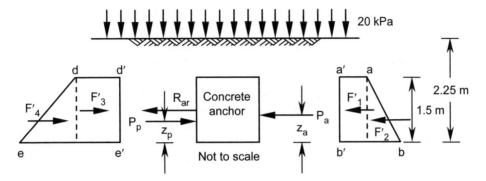

Figure 8.11. Problem 8.10: part (d).

(d) Using Equations 8.2 and 8.4 for layer 1: $k_a = 0.271$, $k_p = 0.369$.
Referring to Figure 8.11:
$a'a = (20.0 + 18.0 \times 0.75) \times 0.271 = 9.1$ kPa.
$b'b = (20.0 + 18.0 \times 2.25) \times 0.271 = 16.4$ kPa.
$P_a = F_1' + F_2' = 9.1 \times 1.5 \times 1.0 + (16.4 - 9.1) \times 1.5 \times 1.0/2 = 13.6 + 5.5 = 19.1$ kN.

$d'd = (20.0 + 18.0 \times 0.75) \times \dfrac{3.69}{1.5} = 82.4$ kPa.

$e'e = (20.0 + 18.0 \times 2.25) \times \dfrac{3.690}{1.5} = 148.8$ kPa.

$P_p = F_3' + F_4' = 82.4 \times 1.5 \times 1.0 + (148.8 - 82.4) \times 1.5 \times 1.0/2 = 123.6 + 49.8 = 173.4$ kN.
For horizontal equilibrium with $s$ being the distance between successive anchor rods:
$R_{ar} \times s + P_a \times 2.0$ (width of the concrete anchor) $-P_p \times 2.0 = 0$.
$119.2 \times s + 19.1 \times 2.0 - 173.4 \times 2.0 = 0 \rightarrow s \approx 2.6$ m.

Problem 8.11

An earth retaining wall 12 m high is reinforced with metal strips of width 100 mm, the
first row of which is at a depth of 0.5 m. The strips are spaced at $s_x = 1$ m and $s_z = 1$ m,
and the allowable tensile strength of the metal strip is 140 MPa. The thickness of the strip
is 5 mm.
The properties of the backfill soil are:
$c' = 0$, $\phi' = 35°$, and $\gamma = 18$ kN/m$^3$.
The friction angle mobilized in the soil-strip contact area is $\delta_b = 23°$.
Calculate the length of the reinforcements at depths 0.5 m, 5.5 m and 11.5 m.

Solution:

The allowable tensile strength of the reinforcement is computed from the following
equation:

$$T_{all} = \frac{T_{ult} \times CRF}{F_c \times F_d \times F_s} \tag{8.36}$$

where $T_{all}$ and $T_{ult}$ represent the allowable and ultimate tensile strengths respectively. The parameter $CRF < 1$ is the creep reduction factor at the order of 40% and higher. The factor of safety of $F_c = 1.3$ to 1.4 is applied for construction damage and $F_s = 1.2$ (slightly higher for retaining walls) is a general factor of safety to take account of the possibility of reaching a limit state and other uncertainties during the design life. If necessary, factors of safety against chemical damages and biological degradations ($F_d$) must be considered. The tensile force in the reinforcement is redistributed in the soil through the surface area of the reinforcement. The friction angle mobilized between the reinforcement and the adjacent soil $\delta_b$ is related to the internal friction angle of soil $\phi'$ using the following equation, where the magnitude of $f_b$ depends on the type of the reinforcement material and varies from 0.5 to 1.0.

$$\tan \delta_b = f_b \tan \phi' \tag{8.37}$$

Depending on the type of the reinforcement, two design methods are available.
1. For reinforcement with high extensibility such as geotextiles, the active state in the earth wall is fully mobilized and the active Rankine earth pressure coefficient $k_a$ is used.
2. For reinforcement of low extensibility (e.g. metal strips), an active zone that is separated from the resisting zone by two linear failure planes (Figure 8.12(a)) are introduced where:

$$H_1 = H + \frac{0.3H \tan \beta}{1 - 0.3 \tan \beta} \tag{8.38}$$

For $\beta = 0$, $H_1 = H$ and the inclined failure plane may be continued up to the upper ground surface or it can be terminated at $0.3H$ (horizontal distance) from the face of the wall. The distribution of the appropriate earth pressure coefficient is shown in Figure 8.12(b). In the absence of hydrostatic water in the soil, the axial tensile load transferred to each horizontal reinforcement element can be calculated as

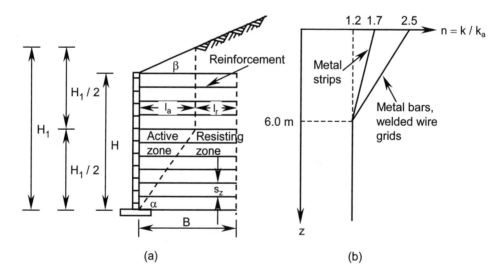

Figure 8.12. Reinforcement of low extensibility: (a) active zone, (b) variation of $n = k / k_a$ with depth.

$$p_r = p_{ah} \times s_x \times s_z = (k_a \sigma_z \cos\beta)s_x s_z \tag{8.39}$$

where $p_r$ is the tensile force in the element, $\sigma_z$ is the vertical stress at depth $z$ within the active zone; $s_x$ and $s_z$ are the horizontal and vertical spacing of the reinforcement elements respectively. The embedment length $l_r$ is calculated to resist the tensile force $p_r$. For a strip element of width $w$, the maximum tensile load $p_t$ due to the frictional bond on both sides is:

$$p_t = 2\sigma_{zr}\cos\beta l_r w \tan\delta_b = 2\sigma_{zr}\cos\beta l_r wf_b \tan\phi' \tag{8.40}$$

where $\sigma_{zr}$ is the vertical stress on the reinforcement element (at depth $z_r$) in the vicinity of the resisting zone. If the ground surface is horizontal and there is no surface loading, or the surface loading is uniform and covers a large area behind the facing panel, we may assume $\sigma_{zr} = \sigma_z$.

Thus, equating Equations 8.39 and 8.40 we obtain:

$$l_r \geq \frac{k_a s_x s_z}{2wf_b \tan\phi'} \tag{8.41}$$

Note that with low extensibility material the term $k_a$ in Equation 8.41 must be replaced by $nk_a$ obtained from Figure 8.12(b). The tensile force in the reinforcement must be equal to or smaller than $T_{all}$ (Equation 8.36). For continuous grid reinforcement, $s_x$ and $w$ are unity and $p_t$ represents the tensile force per unit length of the wall.

Using Equation 8.2 we obtain: $k_a = 0.271$. The position of the potential failure plane is: $\alpha = 45.0° + \phi'/2 = 45.0° + 35.0°/2 = 62.5°$ from the horizontal.

Using Equation 8.41 we have:

$$l_r = \frac{n \times 0.271 \times 1.0 \times 1.0}{2 \times (100.0/1000) \times \tan 23.0°} = 3.19n.$$

Referring to Figure 8.12(b) for metal strips the magnitude of the lateral earth pressure coefficient or the equivalent $n$ value can be found by:

$$\frac{1.7 - 1.2}{0.0 - 6.0} = \frac{n - 1.2}{z - 6.0} \rightarrow n = -\frac{z}{12} + 1.7.$$

At $z = 0.5$ m, $n = 1.66$, at $z = 5.5$ m, $n = 1.24$, and at $z = 11.5$ m, $n = 1.2$.
The corresponding $l_r$ values are:

$l_r$ $(0.5\,\text{m}) = 3.19n = 3.19 \times 1.66 = 5.29$ m,

$l_r$ $(5.5\,\text{m}) = 3.19 \times 1.24 = 3.96$ m, $l_r$ $(11.5\,\text{m}) = 3.19 \times 1.2 = 3.83$ m.

The reinforcement lengths:

$l\,(0.5\,\text{m}) = 5.29 + 0.3H = 5.29 + 0.3 \times 12.0 \approx 8.9$ m,

$l\,(5.5\,\text{m}) = 3.96 + 0.3 \times 12.0 \approx 7.6$ m,

$l\,(11.5\,\text{m}) = 3.83 + (12.0 - 11.5)/\tan 62.5° \approx 4.1$ m.

The axial tensile load from Equation 8.39 (note $\beta = 0$):

$$p_r = p_{ah} \times s_x \times s_z = (k_a \sigma_z \cos\beta)s_x s_z,$$

$$p_r = 0.271 \times n \times 18.0 \times z \times 1.0 \times 1.0 \times 1.0 = 4.878nz.$$

Allowable tesile force $= T_{all} \times$ sectional area of reinforcement element,

Allowable tensile force $= 140.0 \times 1000 \times (100.0/1000) \times (5.0/1000) = 70.0$ kN.

$p_r(0.5\text{m}) = 4.878 \times 1.66 \times 0.5 = 4.0$ kN $< 70.0$ kN.

$p_r(5.5\text{m}) = 4.878 \times 1.24 \times 5.5 = 33.3$ kN $< 70.0$ kN.

$p_r(11.5\text{m}) = 4.878 \times 1.2 \times 11.5 = 67.3$ kN $< 70.0$ kN.

## 8.3   REFERENCES AND RECOMMENDED READINGS

AASHTO 1997. *Standard specifications for highway bridges*. Section 5.8. Washington, D.C.

Alpan, I. 1967. The empirical evaluation of the coefficient $k_o$ and $k_{o,OCR}$. *Soils and foundations*, 7(1): 31-40, Tokyo.

ASTM D-5262. 1997. *Standard test method for evaluating the unconfined tension creep behavior for geosynthetics*. Philadelphia: American Society for Testing and Materials.

Aysen, A. 2002. *Soil mechanics: Basic concepts and engineering applications*. Lisse: Balkema.

Benoit, J. & Lutenegger, A.J. 1993. Determining lateral stress in soft clays. In G.T. Houlsby & A.N. Schofield (eds), *Predictive soil mechanics*. London: Thomas Telford.

Bishop, A.W. 1958. Test requirements for measuring the coefficient of earth pressure at rest. *Proc. Brussels conf. on earth pressure problems*.

Bowles, J.E. 1996. *Foundation analysis and design*. 5th edition. New York: McGraw-Hill.

BS 8006. 1995. *Code of practice for strengthened/reinforced soils and other fills*. London: British Standard Institution.

BS 8081. 1989. *Ground anchorages*. London: British Standard Institution.

Burland, J.B., Potts, D.M. & Walsh, N.M. 1981. The overall stability of free and propped embedded cantilever retaining walls. *Ground engineering*, 14(5): 28-38.

Cheney, R.S. 1988. *Permanent ground anchors*. Report FHWA/DP-68/IR U.S.: 136. Washington, D.C.: Department of Transportation, Federal Highway Administration.

Das, B.M. 1990. *Earth anchors*. Amsterdam: Elsevier.

Das, B.M., Tarquin, A.J. & Moreno, R. 1985. Model tests for pullout resistance of vertical anchors in clay. *Civil eng. for practising and design engineers*, 4(2): 191-209. New York: Program Press.

Duncan, M., Clough, G.W. & Ebeling, R.M. 1990. Behavior and design of gravity earth retaining structures. In P.C. Lambe & L.A. Hansen (eds), *Design and performance of earth retaining structures*: 251-277. New York: ASCE.

Duncan, J.M., & Seed, R.B. 1986. Compaction-induced earth pressure under $k_o$-conditions. *Journal SMFE, ASCE*, 112(1): 1-22.

Exxon. 1992. *Designing for soil reinforcement*. 2nd edition. UK: Exxon Chemical Geopolymers.

Felio, G.Y., Vucetic, M., Hudson, M., Barar, O., & Chapman, R. 1990. Performance of soil nailed walls during the October 17, 1989 Loma Prieta earthquake. *Proc. 43rd Canadian geotechnical conf.*: 165-173. Quebec.

Flaate, K. Peck, R.B. 1973. Braced cuts in sand and clay. *Norwegian geotechnical institute*, Publication 96.

Fleming, W.G.K., Weltman, A.J., Randolph, M.F. & Elson, W.K. 1992. *Piling engineering*. UK: Surrey University Press-Halsted Press.

Hanna, T.H. 1982. *Foundations in tension-ground anchors*. 1st edition: 269-274. USA: McGraw-Hill.

Hueckel, S. 1957. Model tests on anchoring capacity of vertical and inclined plates. *Proc. 4th intern. conf. SMFE*, 2: 203-206. London.

Jáky, J. 1948. Earth pressure in soils. *Proc. 2nd intern. conf. SMFE*, 1: 103-107. Rotterdam.

Juran, I. & Schlosser, F. 1978. Theoretical analysis of failure in reinforced earth structures. *Proc. symp. on earth reinforcement, ASCE convention, Pittsburg*: 528-555. New York: ASCE.

Juran, I., Baudrand, G., Farrag, K., & Elias, V. 1988. Kinematical limit analysis for design of nailed structures. *Journal GED, ASCE*, 116(1): 54-72.

Koerner, R.M. 1994. *Designing with geosynthetics*. 3rd edition. Englewood, New Jersey: Prentice Hall.

Kranz, E. 1953. *Über die verankerung von spundwänden*. Berlin: Verlag Ernst & Sohn.

Kulhawy, F.H., Jackson, C.S. & Mayne, P.W. 1989. First order estimation of $k_o$ in sands and clays. *Foundation engineering: current principles and practices. Journal GED, ASCE*, 1: 121-134.

Littlejohn, G.S. 1970. Soil anchors. *Proc. conf. ground eng.*: 33-44. London: Institution of Civil Engineers:

Littlejohn, G.S. & Bruce, D.A. 1977. *Rock anchors-state of the arts*: 50. UK: Foundation Publications Ltd..

Locher, H.G. 1969. *Anchored retaining walls and cut-off walls*. Berne: Losinger & Co..

Mayne, P.W. & Kulhawy, F.H. 1982. $k_o$-*OCR* relationships in soil. *Journal GED, ASCE*, 108(GT6): 851-872.

Meyerhof, G.G. 1973. Uplift resistance of inclined anchors and piles. *Proc. intern. conf. SMFE*, 2.1: 167-172. Moscow.

Munfakh, G.A. 1990. Innovative earth retaining structures: Selection, design, & performance. In P.C. Lambe & L.A. Hansen (eds), *Design and performance of earth retaining structures*: 85-118. New York: ASCE.

Munfakh, G.A., Abramson, L.W., Barksdale, R.D. & Juran, I. 1987. Soil improvement-a ten year update. In J.P. Welsh (ed.), *ASCE geotechnical special publication*, (12): 59. New York: ASCE.

O'Rourke, T.D. 1987. Lateral stability of compressible walls. *Geotechnique*, 37(1): 145-149.

O'Rourke, T.D. & Jones, C.J.F.P. 1990. Overview of earth retention systems: 1970-1990. In P.C. Lambe & L.A. Hansen (eds), *Design and performance of earth retaining structures*: 22-51. New York: ASCE.

Ostermayer, H. 1977. *Practice on the detail design application of anchorages-A review of diaphragm walls*: 55-61. London: Institution of Civil Engineers.

Ovesen, N.K. & Stromann, H. 1972. Design methods for vertical anchor slabs in sand. *Proc. speciality conf. on performance of earth and earth supported structures*, 2(1): 1481-1500. New York: ASCE.

Pálossy, L., Scharle, P. & Szalatkay, I. 1993. *Earth walls*. New York: Ellis Horwood.

Peck, R.B. 1969. Deep excavation and tunnelling in soft ground. *Proc. 7th intern. conf. SMFE*, 225-290. Mexico.

Peck, R.B. 1990. Fifty years of lateral earth support. In P.C. Lambe & L.A. Hansen (eds), *Design and performance of earth retaining structures*. New York: ASCE.

Post-tensioning Institute (PTI). 1996. *Recommendations for prestressed rock and soil anchors*. Phoenix, Arizona.

Potts, D.M. & Fourie, A.B. 1984. The behaviour of a propped retaining wall: results of a numerical experiment. *Geotechnique*, 34(3): 383-404.

Potts, D.M. & Fourie, A.B. 1985. The effect of wall stiffness on the behaviour of a propped retaining wall. *Geotechnique*, 35(3): 347-352.

Rowe, P.W. 1952. Anchored sheet pile walls. *Proc. institution of civil engineers*, 1(1): 27-70.

Rowe, P.W. 1957. Sheet pile walls in clay. *Proc. institution of civil engineers*, 1(7): 629-654.

Schaefer, V.R., Abramson, L.W., Drumheller, J.C., Hussin, J.D. & Sharp, K.D. 1997. Ground improvement, ground reinforcement, ground treatment-developments 1987-1997. *ASCE geotechnical special publication*, (69): 178-200. New York: ASCE.

Scott, C.R. 1980. *An introduction to soil mechanics and foundations*. 3rd edition. London: Applied Science Publishers.

Task force 27 (Federal Highway Administration). 1990. In-situ soil improvement techniques: design guidelines for use of extensible reinforcements (geosynthetics) for mechanically stabilized earth walls in permanent applications. Joint committee of AASHTO, AGC, ARTBA. Washington, D.C.

Teng, W.C. 1962. *Foundation design*. Englewood Cliffs, New Jersey: Prentice-Hall.

Terzaghi, K.1966. *Theoretical soil mechanics*. 14th edition. New York: John Wiley & Sons.

Terzaghi, K. & Peck, R. B. 1967. *Soil mechanics in engineering practice*. 2nd edition. New York: John Wiley & Sons.

Terzaghi, K., Peck, R. B., & Mesri, G. 1996. *Soil mechanics in engineering practice*. 3rd edition. New York: John Wiley & Sons.

Tschebotarioff, G.P. 1973. *Foundations, retaining and earth structures*. 2nd edition. New York: McGraw-Hill.

Unterreiner, P., Benhamida, B. & Schlosser, F. 1997. Finite element modelling of the construction of a full-scale experimental soil-nailed wall. French national research project CLOUTERRE. *Journal of ground improvement*, 1(1): 1-8.

Williams, B.P. & Waite, D. 1993. *The design and construction of sheet-piled cofferdams*. Special publication 95. London: Construction Industry Research and Information Association.

Williams, G.W., Duncan, J.M. & Sehn, A.L. 1987. Simplified chart solution of compaction-induced earth pressures on rigid structures. *Geotechnical engineering report*. Blackburg, VA.: Virginia Polytechnic Institute and State University.

Wroth, C.P. & Houlsby, G.T. 1985. Soil mechanics-property characterization and analysis. *Proc. intern. conf. SMFE*, 1: 1-56.

Xanthakos, P.P. 1991. *Ground anchors and anchored structures*. New York: John Wiley & Sons.

Xanthakos, P.P., Abramson, L.W. & Bruce, D.A. 1994. *Handbook on ground control and improvement*. New York: John Wiley & Sons.

# CHAPTER 9

# Stability of Earth Slopes

## 9.1   INTRODUCTION

This chapter examines the stability of earth slopes in two-dimensional space using a limit equilibrium method where a mass of soil rotates on a circular or non-circular failure surface at its limit state. On this surface the *Mohr-Coulomb failure criterion* applies and the shear strength parameters used correspond to the peak strength obtained by a total or effective stress analysis. The *factor of safety* is defined as the ratio of the shear strength to the mobilized shear stress on the sliding surface required for equilibrium, and is assumed to be constant along the surface. The average value obtained from a traditional circular analysis is a reliable indication of overall slope stability. Problems 9.1 to 9.3 investigate the slope stability in undrained conditions where the factor of safety can be formulated in terms of geometry of the trial circle and undrained cohesion. For a frictional soil, the mass is divided into vertical slices to facilitate the application of the force and moment equilibrium requirements (Problems 9.4 to 9.7). An infinite slope is normally associated with a translational failure parallel to the ground surface. This case has been explained by Problem 9.8. The circular method can also be applied to evaluate a reinforced slope as shown in Problem 9.9. An alternative method to circular solution is the use of wedge mechanisms for both reinforced soil (Problem 9.10) and unreinforced soil (Problem 9.11).

## 9.2   PROBLEMS

### Problem 9.1

Determine the factor of safety for a 1 vertical to 2 horizontal slope 5 m high using a trial toe circle for which $x_C = 4.5$ m and $y_C = 8$ m (Figure 9.1). The cross-sectional area of the sliding mass is 40.22 m$^2$ and its centroid is located 2.69 m to the right of the centre of the trial circle. The soil properties are:

$c_u = 18$ kPa, $\phi_u = 0$, and $\gamma = 18$ kN/m$^3$.

Solution:

The trial circular failure surface is defined by its centre $C$, radius $R$ and central angle $\theta$. Shear stresses along the trial surface are due only to undrained cohesion $c_u$ and are mobilized to $c_u / F$ (to maintain the equilibrium of the sliding block), where $F$ is the factor of safety. The weight of the sliding block $W$ acts at a distance $d$ from the centre of the circle (Figure 9.1).

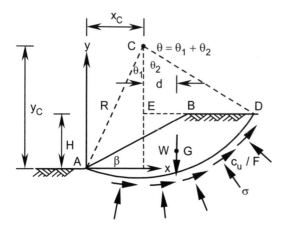

Figure 9.1. Problem 9.1.

Taking moments of the forces about the centre of the circular arc, and noting that the normal stresses on the arc pass through the centre, then:

$$F = \frac{c_u R^2 \theta}{Wd} = \frac{c_u L_a R}{Wd} \qquad (9.1)$$

where $L_a$ is the length of the circular arc.
Geometrical data are:

$$\tan \theta_1 = \frac{x_C}{y_C} = \frac{4.5}{8.0} = 0.5625.$$

$$\theta_1 = 29.4°.$$

$$R = \sqrt{8.0^2 + 4.5^2} = 9.18 \, \text{m}.$$

$$\cos \theta_2 = \frac{8.0 - 5.0}{9.18} = 0.3268.$$

$$\theta_2 = 70.9°.$$

$$\theta = \theta_1 + \theta_2 = 29.36° + 70.92° = 100.3°.$$

$$W = 40.22 \times 1.0 \times 18.0 = 724.0 \, \text{kN}.$$

$$F = \frac{c_u R^2 \theta}{Wd} = \frac{18.0 \times 9.18^2 \times (100.3°/180.0°)\pi}{724.0 \times 2.69} = 1.36.$$

Problem 9.2

For a 45° plane strain slope 30 m high determine:
(a) the factor of safety for a toe circle for which $x_C = 12.5$ m and $y_C = 42$ m if there is no surcharge load on the upper ground surface, and
(b) find the maximum surcharge load $q$ ($L = 20$ m) that will cause the failure of the slope on the same slip circle.
The soil properties are: $c_u = 100$ kPa, $\phi_u = 0$, and $\gamma = 18$ kN/m$^3$.

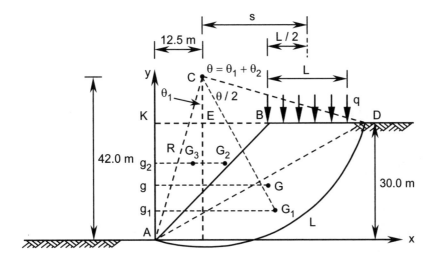

Figure 9.2. Problem 9.2.

Solution:

(a) Area of the sliding mass:

$A = A_1(\text{area of } ADL) + A_2(\text{area of } AKD) - A_3(\text{area of } AKB)$.

$\tan\theta_1 = 12.5/42.0 = 0.2976 \rightarrow \theta_1 = 16.6°$. Radius of the trial circle is:

$R = \sqrt{42.0^2 + 12.5^2} = 43.82$ m.

$\cos\theta_2 = CE/CD = (42.0 - 30.0)/43.82 = 0.2738 \rightarrow \theta_2 = 74.1°$.

$\theta = \theta_1 + \theta_2 = 16.6° + 74.1° = 90.7°$.

Calculation of the area of $ADL$ and the position of its centroid:

$$A_1 = \frac{R^2}{2}(\theta - \sin\theta) = \frac{43.82^2}{2}\left[\pi(90.7°/180.0°) - \sin 90.7°\right] = 559.8 \text{ m}^2.$$

$$C_{G_1} = \frac{4R\ (\sin\theta/2)^3}{3\quad\theta - \sin\theta} = \frac{4\times 43.82}{3}\ \frac{(\sin 90.7°/2)^3}{\pi(90.7°/180.0°) - \sin 90.7°} = 36.08 \text{ m.}$$

$x_{G_1} = g_1 G_1 = x_C + CG_1 \sin(\theta_2 - \theta/2) = 12.5 + 36.08\sin(74.1° - 90.7°/2) = 29.85$ m.

Calculate $A_2$ and $A_3$ and the corresponding $x$ values of the centroids:

$KD = KE + ED = 12.5 + 43.82 \times \sin 74.1° = 54.64$ m.

$A_2 = 30.0 \times 54.64/2 = 819.6 \text{ m}^2.$

$x_{G_2} = g_2 G_2 = KD/3 = 54.64/3 = 18.21$ m.

$A_3 = 30.0 \times 30.0/2 = 450.0 \text{ m}^2.$

$x_{G_3} = g_2 G_3 = KB/3 = 30.0/3 = 10.0$ m.

The area of the sliding mass and the position of its centroid are:

$A = A_1 + A_2 - A_3 = 559.8 + 819.6 - 450.0 = 929.4 \text{ m}^2.$

$$x_G = gG = \frac{x_{G_1} A_1 + x_{G_2} A_2 - x_{G_3} A_3}{A_1 + A_2 - A_3},$$

$$x_G = \frac{29.85 \times 559.8 + 18.21 \times 819.6 - 10.0 \times 450.0}{929.4} = 29.20 \text{ m. Thus}$$

$W = 929.4 \times 18.0 = 16729.2$ kN and:

$d = x_G - x_C = 29.20 - 12.50 = 16.70$ m.

$$F = \frac{c_u R^2 \theta}{Wd} = \frac{100.0 \times 43.82^2 \times (90.7^\circ / 180.0^\circ)\pi}{16729.2 \times 16.70} = 1.09 \approx 1.1.$$

(b) If a surcharge load $q$ is applied to the upper ground surface:

$$F = \frac{c_u R^2 \theta}{Wd + qLs} \tag{9.2}$$

where $L$ is the length of the surcharge load on the plane strain section and $s$ is the horizontal distance of the resultant from the centre of the circle (Figure 9.2).

Check if $L < BD$:

$BD = KD - KB = 54.64 - 30.0 = 24.64$ m.

Thus $L = 20.0$ m $< 24.64$ m.

$s = EB + L/2 = 30.0 - 12.5 + 20.0/2 = 27.5$ m.

$$F = 1 = \frac{c_u R^2 \theta}{Wd + qLs} = \frac{100.0 \times 43.82^2 \times (90.7^\circ / 180.0^\circ)\pi}{16729.2 \times 16.70 + q \times 20.0 \times 27.5},$$

$q = 44.7$ kPa.

Problem 9.3

Using Taylor's stability chart (Figure 9.3), re-solve part (a) of Problem 9.2.

Solution:

The stability of an earth slope in undrained conditions can be expressed in terms of a dimensionless parameter $N$ called the stability number:

$$N = \frac{\gamma H}{c_u} \tag{9.3}$$

For a specified value of $\beta$, the magnitude of $N$ at failure (critical stability number) has a constant value ($N_f$) and the factor of safety (Equation 9.1) may be expressed by:

$$F = \frac{N_f}{N_d} \tag{9.4}$$

where $N_d$ is the stability number corresponding to the design values of $\gamma$, $H$ and $c_u$. In undrained conditions, a horizontal hard stratum located at $n_d H$ below the upper ground surface affects the critical stability number $N_f$. The stability number increases as $n_d$ decreases.

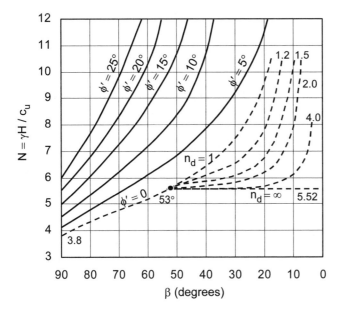

Figure 9.3. Taylor's stability charts used in the solution of Problem 9.3.

For $\beta > 53°$, the critical circle is a toe circle and the hard stratum has no effect on the stability number. The slope stability chart developed by Taylor (1948) is shown in Figure 9.3, where the dashed curves represent the undrained conditions.

For the $c'$, $\phi'$ soils, Taylor adopted a simplifying assumption for the direction of the resultant of the frictional forces acting on the sliding circle, thus a slice method for these soils are more convenient. A slice method facilitates the force equilibrium.

Calculate the stability number corresponding to the design values:

$$N_d = \frac{\gamma H}{c_u} = \frac{18.0 \times 30.0}{100.0} = 5.4.$$

From Figure 9.3 $N_f = 5.52$, therefore:

$$F = \frac{N_f}{N_d} = \frac{5.52}{5.40} = 1.02.$$

Problem 9.4

Using Fellenius' method determine the factor of safety for a slope of 1 vertical to 2 horizontal and height $H = 4.5$ m using a trial toe circle for which $x_C = 4.5$ m and $y_C = 6.25$ m. The soil mass is divided into 4 slices all having identical width of $b = 3$ m, whose average height and angle $\alpha$ are tabulated below.

The soil properties are as follows:

$c' = 6.75$ kPa, $\phi' = 17°$, and $\rho = 1.96$ Mg/m$^3$.

Solution:

| Slice no. | 1 | 2 | 3 | 4 |
|---|---|---|---|---|
| $h$ (m) | 1.6 | 3.7 | 4.6 | 3.0 |
| $\alpha$ (deg.) | −23 | 0 | 23 | 51 |

This method assumes that the shear forces and the normal forces on the two sides of each slice are equal. The factor of safety is expressed by the following equation:

$$F = \frac{\sum_{i=1}^{i=n}[c'l+(w\cos\alpha-ul)\tan\phi']_i}{\sum_{i=1}^{i=n}(w\sin\alpha)_i} = \frac{\sum_{i=1}^{i=n}[c'l+w(\cos\alpha-r_u\sec\alpha)\tan\phi']_i}{\sum_{i=1}^{i=n}(w\sin\alpha)_i} \qquad (9.5)$$

where $l$ is the length of the arc, $w$ is the weight of each slice, $\alpha$ is the angle of the base of the slice from horizontal, $u$ is the pore pressure at the base of each slice, and $n$ is the total number of slices. The pore pressure ratio $r_u$ is a dimensionless parameter defined by:

$$r_u = \frac{ub}{w} \qquad (9.6)$$

which is an alternative replacement for pore pressure $u$.
The results of the calculations are summarized in the table below.

| Slice | $h$ (m) | $\alpha$ (deg.) | $w$ (kN) | $w\cos\alpha$ (kN) | $w\sin\alpha$ (kN) |
|---|---|---|---|---|---|
| 1 | 1.6 | −23 | 92.29 | 84.95 | −36.06 |
| 2 | 3.7 | 0 | 213.43 | 213.43 | 0.00 |
| 3 | 4.6 | 23 | 265.34 | 244.25 | 103.68 |
| 4 | 3.0 | 51 | 173.05 | 108.90 | 134.48 |
| Total: | | | | 651.53 | 202.10 |

Sample calculation for slice 3:
$w = b \times h \times 1.0 \times \rho \times 9.81 = 3.0 \times 4.6 \times 1.96 \times 9.81 = 265.34$ kN.
$w\cos\alpha = 265.34 \times \cos 23.0° = 244.25$ kN.
$w\sin\alpha = 265.34 \times \sin 23.0° = 103.68$ kN.

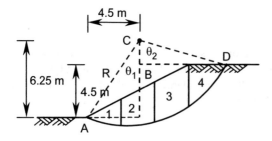

Figure 9.4. Problem 9.4.

From Figure 9.4:

$$R = \sqrt{4.5^2 + 6.25^2} = 7.7 \text{ m.}$$

$$\tan\theta_1 = 4.5/6.25 = 0.72 \rightarrow \theta_1 = 35.75°,$$

$$\cos\theta_2 = (6.25 - 4.5)/7.7 = 0.2273 \rightarrow \theta_2 = 76.86°,$$

$$\theta = \theta_1 + \theta_2 = 35.75° + 76.86° = 112.61°.$$

Using Equation 9.5:

$$F = \frac{\sum\limits_{i=1}^{i=4}\left[c'l + (w\cos\alpha\tan\phi')\right]_i}{\sum\limits_{i=1}^{i=4}(w\sin\alpha)_i} = \frac{c'L_a + \tan\phi'\sum\limits_{i=1}^{i=4}(w\cos\alpha)_i}{\sum\limits_{i=1}^{i=4}(w\sin\alpha)_i},$$

$$F = \frac{6.75 \times 7.7 \times \pi(112.61°/180.0°) + 651.53 \times \tan 17.0°}{202.1} = 1.49.$$

Problem 9.5

A 5 m high slope has an angle of $\beta = 45°$. Data on the 1 m wide slices are given in the table below where $h_w$ is the height of water measured from the mid-point of the base of each slice. The trial circle is not a toe circle and slices 1 and 2 are located to the left of the toe. Using both Fellenius' method, and Bishop's method determine the factor of safety for this trial circle.

$c' = 15$ kPa, $\phi' = 20°$, $\gamma_{sat} = 20.7$ kN/m$^3$, and $\gamma_d = 17.5$ kN/m$^3$.

Solution:

| Slice no. | $h$ (m) | $h_w$ (m) | $\alpha$ (deg.) |
|---|---|---|---|
| 1 | 0.20 | 0.20 | −24.0 |
| 2 | 0.60 | 0.60 | −14.0 |
| 3 | 1.35 | 1.35 | −11.0 |
| 4 | 2.40 | 2.40 | −3.0 |
| 5 | 3.40 | 3.20 | 0.0 |
| 6 | 4.35 | 3.60 | 5.5 |
| 7 | 5.25 | 3.80 | 11.5 |
| 8 | 5.60 | 3.80 | 14.0 |
| 9 | 5.25 | 3.70 | 24.0 |
| 10 | 4.75 | 3.40 | 29.0 |
| 11 | 4.20 | 3.10 | 32.5 |
| 12 | 3.50 | 2.60 | 38.5 |
| 13 | 2.50 | 1.70 | 46.0 |
| 14 | 1.25 | 0.60 | 57.0 |

Sample calculation for slice 8:

$$w = b \times h_w \times \gamma_{sat} + b(h - h_w)\gamma_{dry} = 1.0 \times 3.80 \times 20.7 + 1.0(5.6 - 3.8)17.5 = 110.16 \text{ kN.}$$

$wcos\alpha = 110.16 \times cos14.0° = 106.89$ kN.

$l = b/cos\alpha = 1.0/cos14.0° = 1.031$ m.

$ul = \gamma_w \times h_w \times l = 9.81 \times 3.80 \times 1.031 = 38.43$ kN.

$(wcos\alpha - ul)tan\phi' = (106.89 - 38.43)tan20.0° = 24.92$ kN.

$c' \times l = 15.0 \times 1.031 = 15.46$ kN.

$c'l + (wcos\alpha - ul)tan\phi' = 15.46 + 24.92 = 40.38$ kN.

$wsin\alpha = 110.16 \times sin14.0° = 26.65$ kN.

| Slice no. | 1 | 2 | 3 | 4 | 5 | 6 | 7 | 8 |
|---|---|---|---|---|---|---|---|---|
| | $w$ (kN) | $wcos\alpha$ | $l$ (m) | $ul$ | [(2)–(4)] $tan\phi'$ | $c'l$ | (5) + (6) | $wsin\alpha$ |
| 1 | 4.14 | 3.78 | 1.095 | 2.15 | 0.59 | 16.42 | 17.01 | −1.68 |
| 2 | 12.42 | 12.05 | 1.031 | 6.07 | 2.18 | 15.46 | 17.64 | −3.00 |
| 3 | 27.94 | 27.43 | 1.019 | 13.49 | 5.07 | 15.28 | 20.35 | −5.33 |
| 4 | 49.68 | 49.61 | 1.001 | 23.57 | 9.48 | 15.01 | 24.49 | −2.60 |
| 5 | 69.74 | 69.74 | 1.000 | 31.39 | 13.96 | 15.00 | 28.96 | 0.00 |
| 6 | 87.64 | 87.24 | 1.005 | 35.49 | 18.83 | 15.07 | 33.90 | 8.40 |
| 7 | 104.04 | 101.95 | 1.020 | 38.02 | 23.27 | 15.30 | 38.57 | 20.74 |
| 8 | 110.16 | 106.89 | 1.031 | 38.43 | 24.92 | 15.46 | 40.38 | 26.65 |
| 9 | 103.71 | 94.74 | 1.095 | 39.74 | 20.02 | 16.42 | 36.44 | 42.18 |
| 10 | 94.00 | 82.21 | 1.143 | 38.12 | 16.05 | 17.14 | 33.19 | 45.57 |
| 11 | 83.42 | 70.36 | 1.186 | 36.07 | 12.48 | 17.79 | 30.27 | 44.82 |
| 12 | 69.57 | 54.45 | 1.278 | 32.60 | 7.95 | 19.17 | 27.12 | 43.31 |
| 13 | 49.19 | 34.17 | 1.439 | 24.00 | 3.70 | 21.58 | 25.28 | 35.38 |
| 14 | 23.79 | 12.96 | 1.836 | 10.81 | 0.78 | 27.54 | 28.32 | 19.95 |
| Total: | | | | | | | 401.92 | 274.39 |

Using Equation 9.5:

$$F = \frac{401.92}{274.39} \approx 1.46.$$

The Bishop's method assumes that only the shear forces on the two sides of each slice are equal. This method is considered to be more accurate than Fellenius' method. An increase of 5% to 20% in the factor of safety over Fellenius' method is usually realised. The factor of safety is calculated from:

$$F = \frac{1}{\sum_{i=1}^{i=n}(wsin\alpha)_i}\sum_{i=1}^{i=n}\left[\frac{c'b + w(1-r_u)tan\phi'}{m_\alpha}\right]_i \qquad (9.7)$$

where $m_\alpha$ is defined by:

$$m_\alpha = cos\alpha + \frac{sin\alpha \, tan\phi'}{F} \qquad (9.8)$$

Equation 9.7 is non-linear in $F$ and is solved by fixed-point iteration.

A summary of the computations is tabulated below where 3 iterations have been carried out. The initial value for $F$ in the first iteration was taken 1.5. For the second iteration, the initial value of $F$ is that computed from the first iteration (1.57). This iteration yields $F = 1.57$, thus the initial and computed factors of safety become equal within a computational error less than 0.005. As a further check a third iteration is conducted with an initial value of $F = 1.6$ which gives a smaller value of 1.58 indicating the initial value must decrease. In general the factor of safety is selected as 1.57, which is 7.5% higher than Fellenius' method. Sample calculation for slice 8:

$c'b = 15.0 \times 1.0 = 15.0$ kN.

$w(1-r_u)\tan\phi' = w(1-ub/w)\tan\phi' = (w-ub)\tan\phi',$

$w(1-r_u)\tan\phi' = (110.16 - 9.81 \times 3.8 \times 1.0)\tan 20.0° = 26.53$ kN.

For first iteration with initial $F = 1.5$:

$$m_\alpha = \cos 14.0° + \frac{\sin 14.0° \times \tan 20.0°}{1.5} = 1.029.$$

$$\frac{c'b + w(1-r_u)\tan\phi'}{m_\alpha} = \frac{15.0 + 26.53}{1.029} = 40.36 \text{ kN}.$$

| | (a) | (b) | (c) | F = 1.50 | | F = 1.57 | | F = 1.60 | |
|---|---|---|---|---|---|---|---|---|---|
| Slice | c' b | $w(1-r_u)$ $\tan\phi'$ | $(a) + (b)$ | (d) $\overline{m_\alpha}$ | (c)/(d) | (d) $\overline{m_\alpha}$ | (c)/(d) | (d) $\overline{m_\alpha}$ | (c)/(d) |
| 1 | 15.0 | 0.79 | 15.79 | 0.815 | 19.37 | 0.819 | 19.28 | 0.821 | 19.23 |
| 2 | 15.0 | 2.38 | 17.38 | 0.912 | 19.06 | 0.914 | 19.01 | 0.915 | 18.99 |
| 3 | 15.0 | 5.35 | 20.35 | 0.935 | 21.76 | 0.937 | 21.72 | 0.938 | 21.69 |
| 4 | 15.0 | 9.51 | 24.51 | 0.986 | 24.86 | 0.986 | 24.86 | 0.987 | 24.83 |
| 5 | 15.0 | 13.96 | 28.96 | 1.000 | 28.96 | 1.000 | 28.96 | 1.000 | 28.96 |
| 6 | 15.0 | 19.04 | 34.04 | 1.019 | 33.40 | 1.018 | 33.44 | 1.017 | 33.47 |
| 7 | 15.0 | 24.30 | 39.30 | 1.028 | 38.23 | 1.026 | 38.30 | 1.025 | 38.34 |
| 8 | 15.0 | 26.53 | 41.53 | 1.029 | 40.36 | 1.026 | 40.48 | 1.025 | 40.52 |
| 9 | 15.0 | 24.54 | 39.54 | 1.012 | 39.07 | 1.008 | 39.23 | 1.006 | 39.30 |
| 10 | 15.0 | 22.07 | 37.07 | 0.992 | 37.37 | 0.987 | 37.56 | 0.985 | 37.63 |
| 11 | 15.0 | 19.29 | 34.29 | 0.974 | 35.20 | 0.968 | 35.42 | 0.966 | 35.50 |
| 12 | 15.0 | 16.04 | 31.04 | 0.934 | 33.23 | 0.927 | 33.48 | 0.924 | 33.59 |
| 13 | 15.0 | 11.83 | 26.83 | 0.869 | 30.87 | 0.861 | 31.16 | 0.858 | 31.27 |
| 14 | 15.0 | 6.52 | 21.52 | 0.748 | 28.77 | 0.739 | 29.12 | 0.735 | 29.28 |
| | | | | Total = 430.51 | | Total = 432.02 | | Total = 432.60 | |
| | | From Equation 9.7: | | $F \approx 1.57$ | | $F \approx 1.57$ | | $F \approx 1.58$ | |

Problem 9.6

A 7.1 m high slope for which $\beta = 35°$ is composed of two layers of soil with the following properties (Figure 9.5):

Upper layer: 3.6 m thick, $c'_1 = 20$ kPa, $\phi'_1 = 15°$, $\gamma_1 = 18$ kN/m$^3$.

Lower layer: 3.5 m thick, $c'_2 = 6.3$ kPa, $\phi'_2 = 25°$, $\gamma_2 = 20$ kN/m$^3$.

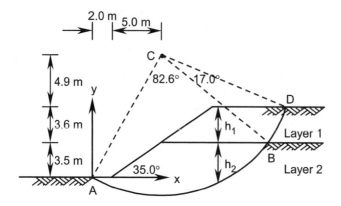

Figure 9.5. Problem 9.6.

The centre of the trial circle is located 5 m to the right and 12 m above the toe and its radius is 13.9 m. Consequently the failure surface is not a toe circle but passes from a point 2 m to the left of the toe. The total central angle is 99.6° for which the central angles corresponding to the upper and lower layers are 17° and 82.6° respectively. Data on the slices are tabulated below. There are 10 slices 2 m wide each. Calculate the weight of each slice from the equation:

$$w = b(\gamma_1 h_1 + \gamma_2 h_2),$$

where $h_1$, $h_2$ are the heights corresponding to layers 1 and 2 respectively. The angle $\alpha$ for each slice is to be computed from:

$$\alpha = \sin^{-1}\frac{\overline{x}}{R},$$

where $\overline{x}$ is the horizontal distance from the mid-point of the base to the centre of the circle. Using Fellenius' method, calculate the factor of safety for the given trial circle. $r_u = 0.3$.

| Slice no. | 1 | 2 | 3 | 4 | 5 | 6 | 7 | 8 | 9 | 10 |
|---|---|---|---|---|---|---|---|---|---|---|
| $h_1$ (m) | 0.0 | 0.0 | 0.0 | 0.0 | 1.4 | 2.8 | 3.5 | 3.6 | 3.6 | 2.4 |
| $h_2$ (m) | 0.5 | 2.0 | 3.8 | 5.4 | 5.2 | 4.8 | 4.0 | 2.8 | 1.6 | 0.0 |

Solution:
The results of calculations are summarized in the table below.
Note that the arc related to the last slice is located in the upper layer:
Equation of the trial circle:

$$(x - x_C)^2 + (y - y_C)^2 = (5.0 + 2.0)^2 + (3.5 + 3.6 + 4.9)^2 = 193.0. \text{ For } y = 3.5 \text{ m:}$$

$$(x_B - 7.0)^2 + (3.5 - 12.0)^2 = 193.0, x_B \approx 18.0 \text{ m.}$$

This means the right edge of slice 9 passes through point $B$ and only the base of slice 10 $(BD)$ is located in layer 1. Slices 1 and 10 have both triangular shape. In a triangular slice it is convenient to calculate $\alpha$ at the intersection point of a vertical line passing through centroid of the triangle with the base. Sample calculations for slice 6:

$w = b(\gamma_1 h_1 + \gamma_2 h_2) = 2.0(18.0 \times 2.80 + 20.0 \times 4.8) = 292.8$ kN.

$\bar{x} = 6b - 0.5b - x_C = 6 \times 2.0 - 0.5 \times 2.0 - 7.0 = 4.0$ m.

$\alpha = \sin^{-1} \dfrac{\bar{x}}{R} = \sin^{-1} \dfrac{4.0}{13.9} = 16.7°$.

$\cos\alpha - r_u \sec\alpha = \cos 16.7° - 0.3 \times \sec 16.7° = 0.645$.

$w(\cos\alpha - r_u \sec\alpha) = 292.8 \times 0.645 = 188.9$ kN.

$w\sin\alpha = 292.8 \times \sin 16.7° = 84.1$ kN.

Calculation of shear resistance due to cohesion:

$\sum c' L_a = 20.0 \times 13.9 \times 17.0° \times \pi/180.0° + 6.3 \times 13.9 \times 82.6° \times \pi/180.0° = 208.7$ kN.

| Slice no. | 1 | 2 | 3 | 4 | 5 | 6 |
|---|---|---|---|---|---|---|
| | $w$ (kN) | $\bar{x}$ (m) | $\alpha$ (deg.) | $\cos\alpha - r_u \sec\alpha$ | (1) × (4) (kN) | $w\sin\alpha$ (kN) |
| 1 | 20.0 | − 5.67 | − 24.1 | 0.584 | 11.7 | − 8.2 |
| 2 | 80.0 | − 4.0 | − 16.7 | 0.645 | 51.6 | − 23.0 |
| 3 | 152.0 | − 2.0 | − 8.3 | 0.686 | 104.3 | − 21.9 |
| 4 | 216.0 | 0.0 | 0.0 | 0.700 | 151.2 | 0.0 |
| 5 | 258.4 | 2.0 | 8.3 | 0.686 | 177.3 | 37.3 |
| 6 | 292.8 | 4.0 | 16.7 | 0.645 | 188.9 | 84.1 |
| 7 | 286.0 | 6.0 | 25.6 | 0.569 | 162.7 | 123.6 |
| 8 | 241.6 | 8.0 | 35.1 | 0.451 | 109.0 | 138.9 |
| 9 | 193.6 | 10.0 | 46.0 | 0.263 | 50.9 | 139.3 |
| Total (1) | | | | | 1007.6 | |
| 10 | 86.4 | 11.67 | 57.1 | − 0.009 | 0.0 | 72.5 |
| Total (2) | | | | | 0.0 | |
| Total: | | | | | 1007.6 | 542.6 |

Using Equation 9.5:

$$F = \frac{\displaystyle\sum_{i=1}^{i=n}\left[c'l + w(\cos\alpha - r_u \sec\alpha)\tan\phi'\right]_i}{\displaystyle\sum_{i=1}^{i=n}(w\sin\alpha)_i} = \frac{208.7 + 0 \times \tan 15.0° + 1007.6 \times \tan 25.0°}{542.6} = 1.25$$

**Problem 9.7**

A slope of 1 vertical to 2 horizontal and height of 7.5 m has the following soil properties: $c' = 15$ kPa, $\phi' = 25°$, and $\gamma = 20$ kN/m$^3$. Using the stability coefficients of Bishop & Morgenstern (1960), compute the factor of safety for $r_u = 0.0, 0.2,$ and $0.4$.

Solution:

Bishop & Morgenstern method uses two stability coefficients, $m$ and $n$, that satisfy the following equation:

$$F = m - nr_u \qquad (9.9)$$

To estimate the factor of safety, the procedure is as follows:

1. Calculate $c' / \gamma H$ from the soil and slope data.

2. For a value of $c' / \gamma H$ just greater than that found in step 1, use the corresponding section of table below and find $m$ and $n$ for $n_d = 1$. Use linear interpolation (for $\phi'$ values) if necessary.

3. If $n$ is underlined the critical circle is at a greater depth. Use the next higher value of $n_d$ to find a non-underlined $n$. Use linear interpolation (for $\phi'$ values) if necessary.

4. Repeat steps 2 and 3 for values of $c' / \gamma H$ just less than that found in step 1.

5. Use Equation 9.9 to obtain two factors of safety for the upper and lower values of $c' / \gamma H$. Calculate the final factor of safety by interpolating between these two values.

$c'/\gamma H = 15.0/(20.0 \times 7.5) = 0.1$.

Thus there is no need to try steps 2 and 4 as Table 9.1 is for $c' / \gamma H = 0.1$.

Table 9.1. Stability coefficients by Bishop & Morgenstern 1960 (for $c'/\gamma H = 0.100$) recalculated by Whitlow (1990).

| $n_d$ | $\phi'$ | cot β 0.5:1 | | 1:1 | | 2:1 | | 3:1 | | 4:1 | |
|---|---|---|---|---|---|---|---|---|---|---|---|
| | | $m$ | $n$ | $m$ | $n$ | $m$ | $n$ | $m$ | $n$ | $m$ | $n$ |
| 1.00 | 20° | 0.98 | 0.80 | 1.25 | 0.86 | 1.83 | <u>1.13</u> | 2.41 | <u>1.46</u> | 2.97 | <u>1.83</u> |
| | 25° | 1.10 | 1.02 | 1.41 | 1.07 | 2.09 | <u>1.42</u> | 2.78 | <u>1.84</u> | 3.36 | <u>2.29</u> |
| | 30° | 1.21 | 1.25 | 1.58 | 1.30 | 2.37 | 1.72 | 3.17 | <u>2.25</u> | 3.91 | <u>2.80</u> |
| | 35° | 1.34 | 1.50 | 1.77 | 1.57 | 2.68 | 2.08 | 3.59 | <u>2.71</u> | 4.49 | <u>3.34</u> |
| | 40° | 1.48 | 1.78 | 1.99 | 1.87 | 3.01 | 2.44 | 4.07 | <u>3.21</u> | 5.10 | <u>3.97</u> |
| 1.25 | 20° | 1.48 | 1.03 | 1.52 | 1.09 | 1.86 | 1.29 | 2.27 | 1.55 | 2.74 | <u>1.83</u> |
| | 25° | 1.72 | 1.29 | 1.79 | 1.38 | **2.19** | **1.63** | 2.67 | 1.96 | 3.21 | <u>2.32</u> |
| | 30° | 1.99 | 1.59 | 2.08 | 1.73 | 2.53 | 2.00 | 3.09 | 2.41 | 3.73 | <u>2.84</u> |
| | 35° | 2.27 | 1.90 | 2.40 | 2.07 | 2.91 | 2.41 | 3.58 | 2.90 | 4.30 | 3.44 |
| | 40° | 2.58 | 2.23 | 2.74 | 2.44 | 3.33 | 2.85 | 4.09 | 3.44 | 4.96 | 4.11 |
| 1.50 | 20° | 1.77 | 1.30 | 1.85 | 1.36 | 2.07 | 1.52 | 2.38 | 1.73 | 2.76 | 2.00 |
| | 25° | 2.11 | 1.66 | 2.20 | 1.72 | 2.47 | 1.93 | 2.83 | 2.21 | 3.28 | 2.53 |
| | 30° | 2.48 | 2.05 | 2.58 | 2.11 | 2.90 | 2.38 | 3.33 | 2.72 | 3.86 | 3.12 |
| | 35° | 2.88 | 2.47 | 2.98 | 2.54 | 3.37 | 2.86 | 3.88 | 3.28 | 4.49 | 3.78 |
| | 40° | 3.33 | 2.94 | 3.45 | 3.03 | 3.90 | 3.42 | 4.49 | 3.92 | 5.21 | 4.51 |

For $c' / \gamma H = 0.1$, cot β $= 2.0$ and $n_d = 1$, $n$ is underlined for the range of $\phi'$ from 20° to 25°, thus select $n_d = 1.25$ for a deeper critical circle.

For $\phi' = 25°$, $m = 2.19$, $n = 1.63$.

For $r_u = 0.0$:

$$F = 2.19 - 1.63 \times 0.0 = 2.19.$$

For $r_u = 0.2$:

$$F = 2.19 - 1.63 \times 0.2 = 1.86.$$

For $r_u = 0.4$:

$$F = 2.19 - 1.63 \times 0.4 = 1.54.$$

## Problem 9.8

A long slope is to be constructed using a material with: $c' = 0$, $\phi' = 35°$, and $\gamma_{sat} = 20$ kN/m$^3$. Determine the critical slope angles ($\beta_c$) for both dry condition and steady state flow parallel to the surface. Calculate the factor of safety for both cases if $\beta = \beta_c / 1.5$.

### Solution:

For $c'$, $\phi'$ soil the factor of safety is defined by the ratio of shear strength to shear stress on a failure plane parallel to the ground surface, then:

$$F = \frac{c'}{\gamma H \sin \beta \cos \beta} + \frac{\tan \phi'}{\tan \beta} \qquad (9.10)$$

The critical height $H_c$ is defined by setting Equation 9.10 to unity:

$$H_c = \frac{c'}{\gamma} \left( \frac{\sec^2 \beta}{\tan \beta - \tan \phi'} \right) \qquad (9.11)$$

For the case where $\beta < \phi'$, the factor of safety is always greater than 1 and is computed from Equation 9.10. This means that there is no limiting value for $H$, and at an infinite depth the factor of safety approaches:

$$F = \frac{\tan \phi'}{\tan \beta} \qquad (9.12)$$

For a granular material with $c' = 0$ and $\beta < \phi'$, the factor of safety is computed from Equation 9.10 (or 9.12). The case where $\beta > \phi'$ and $c' = 0$ is always unstable and cannot be applied to practical situations. This means that the critical value of the slope angle is:

$$\beta_c = \phi' \qquad (9.13)$$

For $c_u$, $\phi_u = 0$ soil:

$$H_c = \frac{c_u}{\gamma \sin \beta \cos \beta} = \frac{2c_u}{\gamma \sin(2\beta)} \qquad (9.14)$$

$$\beta_c = 0.5 \sin^{-1} \frac{2c_u}{\gamma H} \qquad (9.15)$$

The case of steady state flow parallel to the slope angle $\beta$ and with the water table at the ground surface:

$$F = \frac{c'}{\gamma H \sin \beta \cos \beta} + \frac{\tan \phi'}{\tan \beta} - \frac{\gamma_w \tan \phi'}{\gamma \tan \beta} \qquad (9.16)$$

If the water level is at some depth but parallel to the ground surface:

$$F = \frac{c'}{\gamma H \sin \beta \cos \beta} + \frac{\tan \phi'}{\tan \beta} - \frac{\gamma_w h_w \tan \phi'}{\gamma H \tan \beta} \qquad (9.17)$$

where $h_w$ is the height of the water above the base of the slice. If the unit weights of the saturated zone and the zone above the water table are not the same, the term $\gamma H$ in Equation 9.17 must be replaced by $\Sigma \gamma H$. Equation 9.17 may be conveniently presented in the following form:

$$F = \frac{c'}{\gamma H \sin \beta \cos \beta} + \frac{\gamma' \tan \phi'}{\gamma \tan \beta} \qquad (9.18)$$

By setting Equation 9.18 to unity the critical height is calculated as:

$$H_c = \frac{c' \sec^2 \beta}{\gamma \tan \beta - \gamma' \tan \phi'} \qquad (9.19)$$

For the values of $\tan \beta < (\gamma' / \gamma) \tan \phi'$, the factor of safety expressed by Equation 9.18 is always greater than 1.0.

At infinite depth the factor of safety is given by:

$$F = \frac{\gamma' \tan \phi'}{\gamma \tan \beta} \qquad (9.20)$$

From Equation 9.20 we can also calculate the factor of safety for a granular material with $c' = 0$. In this case $\tan \beta$ must be less than $(\gamma' / \gamma) \tan \phi'$, otherwise the slope will not be stable. By setting Equation 9.20 to unity a critical slope angle is defined for granular materials:

$$\beta_c = \tan^{-1} \left( \frac{\gamma' \tan \phi'}{\gamma} \right) \qquad (9.21)$$

For the dry conditions the critical angle is according to Equation 9.13:
$\beta_c = \phi' = 35°$.

For the steady state flow parallel to the surface use Equation 9.21:

$$\beta_c = \tan^{-1} \left( \frac{\gamma' \tan \phi'}{\gamma} \right) = \tan^{-1} \left[ \frac{(20.0 - 9.81) \tan 35.0°}{20.0} \right],$$

$\beta_c = \tan^{-1}(0.357) = 19.6°$.

For $\beta = \beta_c / 1.5 = 35.0° / 1.5 = 23.33°$ and dry conditions use Equation 9.10:

$$F = \frac{c'}{\gamma H \sin \beta \cos \beta} + \frac{\tan \phi'}{\tan \beta} = \frac{\tan 35.0°}{\tan 23.33°} = 1.62.$$

For $\beta = 19.6° / 1.5 \approx 13.0°$ and the steady state flow conditions use Equation 9.16:

$$F = \frac{c'}{\gamma H \sin \beta \cos \beta} + \frac{\tan \phi'}{\tan \beta} - \frac{\gamma_w \tan \phi'}{\gamma \tan \beta} = \frac{\tan 35.0°}{\tan 13.0°} - \frac{9.81 \times \tan 35.0°}{20.0 \times \tan 13.0°} = 1.54.$$

Problem 9.9

A 7.2 m high slope, which has a batter of 1.0 horizontal to 1.8 vertical, is to be reinforced with horizontal geosynthetic elements. Properties of the soil are:

$c' = 0$, $\phi' = 35°$, $\gamma = 19$ kN/m$^3$ and $r_u = 0.4$. For a toe circle of radius 10.54 m tangent to the base at the toe compute the total tensile force in the reinforcement assuming a factor of safety of 1.4 and using Bishop's simplified method.

Stability of Earth Slopes 153

Relevant data are given in the table below and $b = 2$ m for all slices.

| Slice no. | 1 | 2 | 3 | 4 | 5 |
|---|---|---|---|---|---|
| $h$ (m) | 1.7 | 4.9 | 5.9 | 4.4 | 1.8 |
| $\alpha$ (deg.) | 5.4 | 16.6 | 28.4 | 41.8 | 59.5 |

Solution:
In the case of horizontal reinforcement in a $c_u$, $\phi_u = 0$ soil, adding the moments of the reinforcement tensile forces to the resisting moments results in the following equation:

$$F = \frac{c_u L_a R + \sum_{j=1}^{j=m}[T(y_C - y)]_j}{Wd} \quad (9.22)$$

where $T_j$ is the reinforcement tensile force, $y_j$ is the vertical distance of the reinforcement from the x-axis and $m$ is the number of the reinforcement layers.
For the slope in the $c'$, $\phi'$ soil Bishop's simplified method is modified as follows:

$$F = \frac{\sum_{i=1}^{i=n}\{[c'b + w(1-r_u)\tan\phi']/m_\alpha\}_i + \frac{1}{R}\sum_{j=1}^{j=m}[T(y_C - y)]_j}{\sum_{i=1}^{i=n}(w\sin\alpha)_i} \quad (9.23)$$

The total reinforcement force $T_{total}$ is the sum of the tensile forces in the reinforcement; it is equivalent to the integral of the lateral soil pressure $\sigma_h$ with the pressure coefficient $k$ less than $k_a$ (for $F = 1$) due to the slope angle. It is convenient to assume a linear distribution for lateral stress with depth. Equations 9.22 and 9.23 are solved for a specified $F$ to yield $T_{total}$:

$$\Delta M = \sum_{j=1}^{j=m}[T(y_C - y)]_j = T_{total}(y_C - H/3) = (\gamma H^2 k/2)(y_C - H/3) \quad (9.24)$$

The results of computations are summarized in the table below.

| Slice | $h$ (m) | $\alpha$ (deg.) | $w$ (kN) | $w\sin\alpha$ (kN) | $w(1 - r_u)\tan\phi' / m_\alpha$ (kN) |
|---|---|---|---|---|---|
| 1 | 1.7 | 5.4 | 64.6 | 6.1 | 26.0 |
| 2 | 4.9 | 16.6 | 186.2 | 53.2 | 71.0 |
| 3 | 5.9 | 28.4 | 224.2 | 106.6 | 84.3 |
| 4 | 4.4 | 41.8 | 167.2 | 111.4 | 65.1 |
| 5 | 1.8 | 59.5 | 68.4 | 58.9 | 30.6 |
| Total: | | | | 336.2 | 277.0 |

Using Equation 9.23:

$$F = 1.4 = \frac{277.0 + \Delta M / 10.54}{336.2} \rightarrow \Delta M = 2041.4 \text{ kN.m.}$$

From Equation 9.24:

$$\Delta M = 2041.4 = T_{total}\left(10.54 - \frac{7.2}{3}\right),$$

$$T_{total} = \frac{2041.4}{8.14} = 250.8 \text{ kN.}$$

Problem 9.10

For the reinforced slope shown in Figure 9.6(a), calculate the total force in the reinforcement for the trial two-part wedge shown. $BC$ is parallel to $AD$.

The soil properties are: $c' = 0$, $\phi' = 29°$, and $\gamma = 18$ kN/m$^3$.

Solution:

With a vertical inter-wedge boundary, the mechanism is defined by three independent variables $h$, $\theta_1$ and $\theta_2$ (Figure 9.6(a)). If the inter-wedge boundary is not vertical then an additional variable is needed to specify the boundary. In the evaluation of stability, only force equilibrium is used. The failure criterion is assumed to apply on the three sliding surfaces, which in turn implies that the shear strength on these surfaces is fully mobilized. In the two-part wedge method adopted by the Department of Transport, UK, it is assumed that the friction angle on the inter-wedge sliding surface is zero; which results in reduced computational effort. Free body diagrams of the wedges (with the above simplifying assumption) are shown in Figure 9.6(b) where the total reinforcement force $T_{total}$ is the sum of $T_1$and $T_2$ corresponding to wedges 1 and 2 respectively.

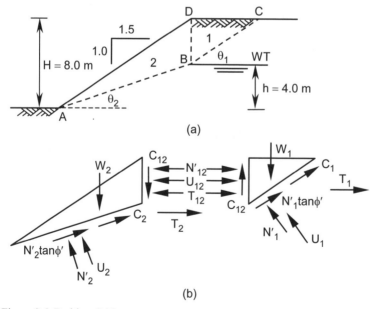

Figure 9.6. Problem 9.10.

The force system is statically determinate and $T_{total}$ can be formulated by considering horizontal and vertical equilibrium of both wedges:

$$T_{total} = \frac{W_1(\tan\theta_1 - \tan\phi') + (U_1\tan\phi' - C_1)\sec\theta_1}{1 + \tan\theta_1\tan\phi'} +$$

$$\frac{W_2(\tan\theta_2 - \tan\phi') + (U_2\tan\phi' - C_2)\sec\theta_2}{1 + \tan\theta_2\tan\phi'} \qquad (9.25)$$

where the first term represents $T_1$, the second term is $T_2$, and $W_1$, $W_2$ are the weights of the wedges 1 and 2, $C_1$, $C_2$ are the forces due to cohesion acting on the sliding bases, and $U_1$, $U_2$ are the forces due to water pressure acting on the sliding bases. Similar to the circular method, the computed value of $T_{total}$ is assumed to be linearly distributed along the slope height. From the geometry of Figure 9.6(a):
$BD = 4.0$ m, $DC = 6.0$ m; the horizontal distance of point $A$ from $BD$ is 12.0 m.

$$\theta_1 = \tan^{-1}\frac{4.0}{6.0} = 33.69° \text{ and,}$$

$$\theta_2 = \tan^{-1}\frac{4.0}{12.0} = 18.43°.$$

Calculate the weights of the wedges:

$$W_1 = \frac{4.0 \times 6.0 \times 1.0}{2} \times 18.0 = 216.0 \text{ kN,}$$

$$W_2 = \frac{4.0 \times 12.0 \times 1.0}{2} \times 18.0 = 432.0 \text{ kN.}$$

As the soil is cohesionless, therefore $C_1 = C_2 = 0$.
Calculate the forces due to water pressure ($U_1$ and $U_2$) on the sliding bases of $BC$ and $AB$:

$$U_1 = 0, \ U_2 = \frac{u_A + u_B}{2} \times 1.0 \times AB.$$

Where $u_A$ and $u_B$ are pore pressures at points $A$ and $B$ respectively.

$$u_B = 0, u_A = 9.81 \times 4.0 = 39.24 \text{ kPa and,}$$

$$AB = \sqrt{4.0^2 + 12.0^2} = 12.65 \text{ m, thus}$$

$$U_2 = \frac{39.24 + 0.00}{2} \times 12.65 = 248.19 \text{ kN.}$$

Substituting the above values in Equation 9.25:

$$T_{total} = \frac{216.0(\tan 33.69° - \tan 29.0°)}{1 + \tan 33.69° \tan 29.0°} +$$

$$\frac{432.0(\tan 18.43° - \tan 29.0°) + (248.19\tan 29.0°)\sec 18.43°}{1 + \tan 18.43° \tan 29.0°} =$$

$T_1 + T_2 = 17.72 + 41.79 = 59.5 \text{kN}.$

Equivalent earth pressure coefficient (assuming a linear earth pressure distribution) from Equation 8.7 (with horizontal upper ground surface) is:

$$T_{total} = 59.5 = \frac{1}{2}\gamma H^2 k = \frac{1}{2} \times 18.0 \times 8.0^2 k.$$

$k = 0.1.$

## Problem 9.11

For the multiple-wedge mechanism shown in Figure 9.7(a), calculate the factor of safety for the slope assuming that no cohesion or friction is mobilized on the vertical inter-wedge planes of $CE$ and $BF$.

$c' = 9.5 \text{ kPa}, \phi' = 30.8°, \text{ and } \gamma = 18 \text{ kN/m}^3.$

Solution:

The free body diagrams of the wedges are shown in Figure 9.7(b) where $c'_m$ and $\phi'_m$ are the mobilized cohesion and friction angle on the non-vertical sliding surfaces. It is assumed no cohesion and friction are mobilized on the vertical interfaces, which means the corresponding internal forces are normal to the boundaries. The solution procedure involves a selection of a factor of safety and investigating the force equilibrium. The internal force $E_2$ calculated from wedge 2 ($E_{2R}$) is compared with the corresponding force from wedge 3 ($E_{2L}$). For $F = 1.9$ these values become equal (within an accepted error). The results of the calculations are tabulated and sample calculation for $F = 1.9$ is included.

For $F = 1.9$:

$$c'_m = \frac{c'}{1.9} = \frac{9.5}{1.9} = 5.0 \text{ kPa}.$$

$$\phi'_m = \tan^{-1}(\frac{\tan \phi'}{1.9}),$$

$$\phi'_m = \tan^{-1}(\frac{\tan 30.8°}{1.9}) = 17.42°.$$

From the geometry of the mechanism it can be shown that:

$EC = 2.0 \text{ m}, ED = 1.40 \text{ m}, CD = 2.44 \text{ m},$

$FE = BC = 11.31 \text{ m}, AF = 2.86 \text{ m}, BA = 3.49 \text{ m}.$

The areas of the blocks are as follows:

$S_1 (CED) = 1.40 \text{ m}^2,$

$S_2 (BFEC) = 16.00 \text{ m}^2,$

$S_3 (AFB) = 2.86 \text{ m}^2.$

Calculate weights and cohesion forces:

$W_1 = 1.40 \times 1.0 \times 18.0 = 25.2 \text{ kN}.$

$W_2 = 16.0 \times 1.0 \times 18.0 = 288.0 \text{ kN}.$

$W_3 = 2.86 \times 1.0 \times 18.0 = 51.48 \text{ kN}.$

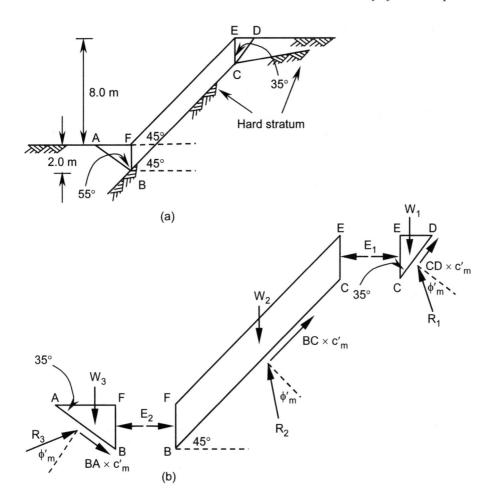

Figure 9.7. Problem 9.11.

Cohesion force along $CD$:

$C_1 = CD \times 1.0 \times c'_m = 2.44 \times 5.0 = 12.2$ kN.

Cohesion force along $BC$:

$C_2 = BC \times 1.0 \times c'_m = 11.31 \times 5.0 = 56.55$ kN.

Cohesion force along $BA$:

$C_3 = BA \times 1.0 \times c'_m = 3.49 \times 5.0 = 17.45$ kN.

Vertical and horizontal equilibrium of wedge $CED$:

$W_1 - C_1 \cos 35.0° - R_1 \cos(55.0° - 17.42°) = 0,$

$25.2 - 12.2 \cos 35.0° - R_1 \cos 37.58° = 0,$

$R_1 = 19.19$ kN.

$E_1 + C_1 \sin 35.0° - R_1 \sin(55.0° - 17.42°) = 0,$

$E_1 + 12.2\sin 35.0° - 19.19\sin 37.58° = 0,$

$E_1 = 4.70$ kN.

Vertical and horizontal equilibrium of wedge *BFEC*:

$W_2 - C_2\cos 45.0° - R_2\cos(45.0° - 17.42°) = 0,$

$288.0 - 56.55\cos 45.0° - R_2\cos 27.58° = 0,$

$R_2 = 279.81$ kN.

$E_{2R} - E_1 + C_2\sin 45.0° - R_2\sin(45.0° - 17.42°) = 0,$

$E_{2R} - 4.7 + 56.55\sin 45.0° - 279.8\sin 27.58° = 0,$

$E_{2R} = 94.26$ kN.

Vertical and horizontal equilibrium of wedge *AFB*:

$W_3 + C_3\cos 55.0° - R_3\cos(35.0° + 17.42°) = 0,$

$51.48 + 17.45\cos 55.0° - R_3\cos 52.42°,$

$R_3 = 100.82$ kN.

$E_{2L} - C_3\sin 55.0° - R_3\sin(35.0° + 17.42°) = 0,$

$E_{2L} - 17.45\sin 55.0° - 100.82\sin 52.42° = 0,$

$E_{2L} = 94.19$ kN $\approx E_{2R} = 94.26$ kN.

| $F$ | $c'_m$ (kPa) | $\phi'_m$ (deg.) | $R_1$ (kN) | $R_2$ (kN) | $R_3$ (kN) | $E_{2R}$ (kN) | $E_{2L}$ (kN) |
|-----|------|-------|-------|--------|--------|-------|--------|
| 1.7 | 5.59 | 19.32 | 17.27 | 269.96 | 107.45 | 74.53 | 103.26 |
| 1.8 | 5.28 | 18.32 | 18.27 | 275.06 | 103.88 | 84.80 | 98.41 |
| 1.9 | 5.00 | 17.42 | 19.19 | 279.81 | 100.82 | 94.26 | 94.19 |

Note that the variables in the geometry of the wedges are not optimised yet. This means with the variation of the angles in wedges 1 and 3 we may obtain a lower value for *F*.

## 9.3  REFERENCES AND RECOMMENDED READINGS

Atkinson, J.H. 1993. *An introduction to the mechanics of soils and foundations.* London: McGraw-Hill.

Aysen, A. 2002. *Soil mechanics: Basic concepts and engineering applications.* Lisse: Balkema.

Aysen, A. & Sloan, S.W. 1992. Stability of slopes in cohesive frictional soil. *Proc. 6th Australia-New Zealand conf. on geomechanics*: Geotechnical risk-identification, evaluation and solutions: 414-419. New Zealand: New Zealand Geomechanics Society.

Aysen, A. & Loadwick, F. 1995. Stability of slopes in cohesive frictional soil using upper bound collapse mechanisms and numerical methods. *Proc. 14th Australasian conf. on the mechanics of structures and materials*, 1: 55-59. Hobart, Australia: University of Hobart.

Bishop, A.W. 1955. The use of slip circle in the stability analysis of slopes. *Geotechnique*, 5(1): 7-17.

Bishop, A.W. & Morgenstern, N.R. 1960. Stability coefficients for earth slopes. *Geotechnique*, 10(4): 129-147.

Bromhead, E.N. 1992. *The stability of slopes*. 2nd edition. Surrey: Surrey University Press.

Celestino, T.B. & Duncan, J.M. 1981. Simplified search for non-circular slip surface. *Proc. 10th intern. conf. SMFE*, 3: 391-394. Rotterdam: Balkema.

Chandler, R.J. & Peiris, T.A.1989. Further extensions to the Bishop and Morgenstern slope stability charts. *Ground engineering*, May: 74-91.

Chen, R.H. & Chameau, J.L. 1982. Three dimensional slope stability analysis. *Proc. 4th intern. conf. numer. meth. in geomech.*, 2: 671-677. Rotterdam: Balkema.

Chen, R.H. & Chameau, J.L. 1983. Three dimensional limit equilibrium analysis of slopes. *Geotechnique*, 33(1): 31-40.

Chen, Z. & Shao, C. 1988. Evaluation of minimum factor of safety in slope stability analysis. *Canadian geotechnical journal*, 25(4): 735-748.

Cousins, B.F. 1978. Stability charts for simple earth slopes. *Journal GE, ASCE*, 104.

Duncan, J.M. 1996. State of the art: Limit equilibrium and finite element analysis of slopes. *Journal GE, ASCE*, 122(7): 577-596.

Duncan, J.M., Buchignani, A.L. & De Wet, M. 1987. *An engineering manual for slope stability studies*. Blacksburg, VA: Virginia Tech.

Exxon Chemicals. 1992. *Geotextiles: design for soil reinforcement*. 2nd edition: 58. UK: Exxon Chemicals Geopolymers Ltd.

Fellenius, W. 1927. *Erdstatische berechnungen mit reibung und kohasion*, (in German). Berlin: Ernst.

Hovland, H.J. 1977. Three dimensional slope stability analysis method. *Journal GE, ASCE*, 103(9): 971-986.

Hunger, O. 1987. An extension of Bishop's simplified method of slope stability analysis to three dimensions. *Geotechnique*, 37(1): 113-117.

Janbu, N. 1968. Slope stability computations. *Soil mechanics and foundation engineering report*. Trondheim, Norway: The Technical University of Norway.

Janbu, N. 1973. Slope stability computations. In E. Hirschfield & S. Poulos (eds), *Embankment dam engineering, Casagrande memorial volume*: 47-86. New York: John Wiley.

Jewell, R.A. 1991. Application of revised design charts for steep reinforced slopes. *Geotextiles & geomembranes*, 10(3): 203-233. UK: Elsevier.

Jewell, R.A., Paine, N. & Woods, R.I. 1985. *Design methods for steep reinforced embankments, polymer grid reinforcement*: 70-81. UK: Thomas Telford.

King, C.J.W. 1989. Revision of effective stress method of slices. *Geotechnique*, 39(3): 497-502.

Koerner, R.M. 1984. Slope stabilization using anchored geotextiles: Anchored spider netting. *Proc. special geotechnical engineering for roads and bridges conf.*: 1-11. Harrisburg, PA: Penn DOT.

Koerner, R.M. & Robins, J.C. 1986. In-situ stabilization of soil slopes using nailed geosynthetics. *Proc. 3rd conf. on geosynthetics*: 395-399. Vienna.

Ladd, C.C. 1991. Stability evaluation during staged construction. *Journal GE, ASCE*, 117(4): 537-615.

Leshchinsky, D. & Huang, C. 1992. Generalized three dimensional slope stability analysis. *Journal GE, ASCE*, 118(11): 1748-1764.

Morgenstern, N.R. & Price, V.E. 1965. The analysis of the stability of general slip surfaces. *Geotechnique*, 15(1): 79-93.

Naylor, D.J. 1991. Finite element methods for fills and embankment dams. In M. das Neves (ed.), *Advances in rockfill structures*. North Atlantic Treaty Organization advanced study institute series: 291-339. Dordrecht, The Netherlands: Kluver Academic Publisher.

Spencer, E. 1967. A method of analysis of the stability of embankments assuming parallel interslice forces. *Geotechnique*, 17(1): 11-26.

Tavenas, F., Trak, B. & Leroueil, S. 1980. Remarks on the validity of stability analyses. *Canadian geotechnical journal*, 17(1): 61-73.

Taylor, D.W. 1948. *Fundamentals of soil mechanics*. New York: Wiley.

UK Department of Transport. 1994. Design methods for the reinforcement of highway slopes by reinforced soil and soil nailing techniques. *Design manual 4, section 1, HA 68/94*.

Whitlow, R. 1990. *Basic soil mechanics*. 2nd edition. New York: Longman Scientific & Technical.

# CHAPTER 10

# Bearing Capacity of Shallow Foundations and Piles

## 10.1 INTRODUCTION

The problems solved in this chapter are related to the *ultimate bearing capacity* of shallow footings, piles and pile groups. The ultimate bearing capacity of a shallow footing is evaluated using traditional methods (Terzaghi, 1943; Meyerhof, 1951, 1953, 1963, 1965 and 1976) including methods by Hansen (1961 and 1970) and Vesić (1973) with modifications by Bowles (1996); (Problems 10.1 to 10.4). All the above solutions introduce *bearing capacity factors* of $N_c$, $N_q$, $N_\gamma$, that are functions of the internal friction angle $\phi'$. These factors represent the effects of cohesion, surcharge load adjacent to the footing and the weight of the soil within the failure zone. The improvements suggested by Hansen and Vesić take account of the geometry of the footing and inclinations of the load and ground surface. The *base capacity* of a pile in $c'$, $\phi'$ soil or $c_u$, $\phi_u = 0$ soil is calculated using modified methods of Hansen and Vesić (Problems 10.5 and 10.6). For piles in sands ($c' = 0$, $\phi'$) a solution proposed by Fleming et al. (1992) is used (Problem 10.7). This solution is based on the known values of density index $I_D$, the critical friction angle $\phi'_{cr}$ and the effective overburden pressure $p'_o$. The $\phi'$-$N_q$ relationship used is taken from the theory of ultimate bearing capacity developed by Berezantzev et al. (1961). The *shaft capacity* of a pile may be calculated by $\alpha$ method (Skempton, 1959; Tomilinson, 1977) where the average limiting shear stress $\tau_s$ mobilized on the shaft is estimated as a fraction of the undrained cohesion $c_u$. The improvements proposed by Fleming et al. (1992) and Randolph and Murphy (1985) are used in the solution of Problem 10.8. An alternative effective stress analysis suggested by Burland (1973), ($\beta$ method) is considered in Problem 10.9. A pile group is treated using Equation 5.50 assuming a rigid cap resting on piles. Thus the individual axial force can be evaluated (Problem 10.10). The settlement of each pile is estimated by establishing the vertical stress distribution under the pile using Mindlin solution (Problem 10.11). With the settlement calculated, The Winkler spring model can be applied to analyse the cap as an elastic beam supported by springs.

## 10.2 PROBLEMS

Problem 10.1

A square footing of 1 m is located at a depth 1.5 m below the ground surface. The soil properties are:

$c' = 0$, $\phi' = 40°$, $\gamma = 16.7$ kN/m$^3$, $\gamma_{sat} = 20$ kN/m$^3$.

Using Terzaghi's bearing capacity factors calculate the ultimate bearing capacity:
(a) The water table is well below the foundation level,
(b) the water table is at the ground surface.

Solution:

The ultimate bearing capacity for a shallow strip, square and circular footings according to Terzaghi:

$$q_u = c'N_c + \gamma DN_q + 0.5B\gamma N_\gamma \tag{10.1}$$

$$q_u = 1.3c'N_c + \gamma DN_q + 0.4B\gamma N_\gamma \tag{10.2}$$

$$q_u = 1.2c'N_c + \gamma DN_q + 0.3B\gamma N_\gamma \tag{10.3}$$

The corresponding bearing capacity factors are:

$$N_q = \frac{e^{(3\pi/2-\phi')\tan\phi'}}{2\cos^2(45°+\phi'/2)},$$

$$N_c = \cot\phi'(N_q - 1),$$

$$N_\gamma = 0.5\tan\phi'(\frac{k_{p\gamma}}{\cos^2\phi'} - 1) \tag{10.4}$$

From the given values of $N_\gamma$ the following matching empirical equation is proposed:

$$k_{p\gamma} = (8\phi'^2 - 4\phi' + 3.8)\tan^2(60° + \phi'/2) \tag{10.5}$$

where $\phi'$ (in the first term) is in radians. Figure 10.1 shows the variation of the bearing capacity factors with the effective internal friction angle $\phi'$.
In the undrained conditions with $c_u$ and $\phi_u = 0$:

$$N_c = \frac{3}{2}\pi + 1 = 5.71, \ N_q = 1 \text{ and } N_\gamma = 0 \tag{10.6}$$

When the water table is on the ground surface the unit weight is replaced by effective unit weight: $\gamma_{sat} - \gamma_w$, where $\gamma_w$ is the unit weight of water. If the water table is below the ground surface then a linear interpolation can be adopted between two results one with the saturated unit weight and one with the unit weight above the water table.
Using Equations 10.4:

$$N_q = \frac{e^{(3\pi/2-40.0°\times\pi/180.0°)\tan 40.0°}}{2\cos^2(45.0°+40.0°/2)} = 81.27.$$

As the soil underneath of the footing is cohesionless, there is no need to calculate $N_c$.

$$k_{p\gamma} = \left[8(\frac{40.0°\times\pi}{180.0°})^2 - 4\times\frac{40.0°\times\pi}{180.0°} + 3.8\right]\tan^2(60.0° + 40.0°/2) = 157.81.$$

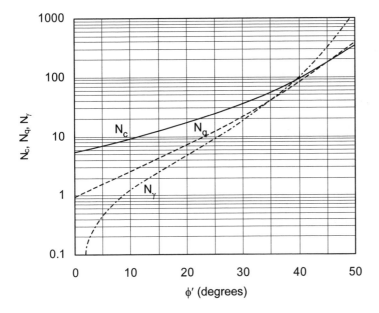

Figure 10.1. Bearing capacity factors using Terzaghi's equations.

The proposed equation for $k_{p\gamma}$ overestimates the $N_\gamma$ value about 3% to 5% (at high $\phi'$ values) which is not significant considering that the Terzaghi's method underestimates the ultimate bearing capacity nearly by a factor of 1.5.

$$N_\gamma = 0.5 \times \tan 40.0°(\frac{157.81}{\cos^2 40.0°} - 1) = 112.41.$$

The Terzaghi's value for $N_\gamma$ is 100.39.

(a) $q_u = c'N_c + \gamma DN_q + 0.5 B\gamma N_\gamma = 16.7 \times 1.5 \times 81.27 + 0.4 \times 1.0 \times 16.7 \times 112.41 = 2787$ kPa.

Using $N_\gamma = 100.39$, $q_u = 2706$ kPa.

(b) In this case we use the effective unit weight (or submerged unit weight):

$$\gamma' = \gamma_{sat} - \gamma_w = 20.0 - 9.81 = 10.19 \text{ kN/m}^3.$$

$q_u = 10.19 \times 1.5 \times 81.27 + 0.4 \times 1.0 \times 10.19 \times 112.41 = 1700$ kPa.

Using $N_\gamma = 100.39$, $q_u = 1651$ kPa.

Problem 10.2

Re-work Problem 10.1 using Meyerhof's bearing capacity equations.

Solution:

In the Meyerhof's bearing capacity equations the shape of the footing, inclination of the applied load and the depth of the footing are taken into account by introducing the corresponding factors of $s$, $i$, and $d$.

For a rectangular footing of $L$ by $B$ ($L > B$):

$$q_u = c'N_c s_c i_c d_c + \gamma D N_q s_q i_q d_q + 0.5 B \gamma N_\gamma s_\gamma i_\gamma d_\gamma \tag{10.7}$$

For vertical load: $i_c = i_q = i_\gamma = 1$ and:

$$q_u = c'N_c s_c d_c + \gamma D N_q s_q d_q + 0.5 B \gamma N_\gamma s_\gamma d_\gamma \tag{10.8}$$

The bearing capacity factors are:

$$N_q = \exp(\pi \tan \phi') \tan^2(45° + \phi'/2), \ N_c = \cot \phi'(N_q - 1),$$

$$N_\gamma = (N_q - 1)\tan(1.4\phi') \tag{10.9}$$

The shape, inclination and depth factors are according:

$$s_c = 1 + 0.2\frac{B}{L}\tan^2(45° + \phi'/2),$$

$$s_q = s_\gamma = 1 + 0.1\frac{B}{L}\tan^2(45° + \phi'/2) \tag{10.10}$$

For $c_u$, $\phi_u = 0$ soil $s_q = s_\gamma = 1$.

$$i_c = i_q = \left(1 - \frac{\alpha°}{90°}\right)^2, \ i_\gamma = \left(1 - \frac{\alpha}{\phi'}\right)^2 \tag{10.11}$$

For $c_u$, $\phi_u = 0$ soil $i_\gamma = 1$.

$$d_c = 1 + 0.2\frac{D}{B}\tan(45° + \phi'/2),$$

$$d_q = d_\gamma = 1 + 0.1\frac{D}{B}\tan(45° + \phi'/2) \tag{10.12}$$

For $c_u$, $\phi_u = 0$ soil $d_q = d_\gamma = 1$. The equivalent plane strain $\phi'$ is related to triaxial $\phi'$ by:

$$\phi'_{ps} = \phi'_{tri}\left(1.1 - 0.1\frac{B}{L}\right) \tag{10.13}$$

For the eccentric load the length and width of the rectangular footing are modified to:

$$L' = L - 2e_L, B' = B - 2e_B \tag{10.14}$$

where $e_L$ and $e_B$ represent the eccentricity along the appropriate directions.
From Equations 10.9:

$$N_q = e^{\pi \tan 40.0°}\tan^2(45.0° + 40.0°/2) = 64.19.$$
$$N_\gamma = (64.19 - 1)\tan(1.4 \times 40.0°) = 93.68.$$

Using Equations 10.10 and 10.12:

$$s_q = s_\gamma = 1 + 0.1\frac{B}{L}\tan^2(45° + \phi'/2) = 1 + 0.1\frac{1.0}{1.0}\tan^2(45.0° + 40.0°/2) = 1.46.$$

$$d_q = d_\gamma = 1 + 0.1\frac{D}{B}\tan(45° + \phi'/2) = 1.0 + 0.1\frac{1.5}{1.0}\tan(45.0° + 40.0°/2) = 1.32.$$

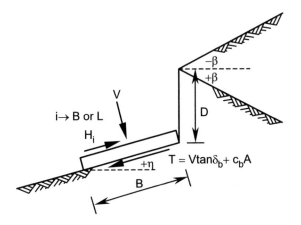

Figure 10.2. Identification of the terms in Hansen's bearing capacity equations.

From Equation 10.8:

(a) $q_u = 16.7 \times 1.5 \times 64.19 \times 1.46 \times 1.32 + 0.5 \times 1.0 \times 16.7 \times 93.68 \times 1.46 \times 1.32 = 4606\,\text{kPa}$.

(b) $q_u = 10.19 \times 1.5 \times 64.19 \times 1.46 \times 1.32 + 0.5 \times 1.0 \times 10.19 \times 93.68 \times 1.46 \times 1.32 = 2811\,\text{kPa}$.

Problem 10.3

Re-work Problem 10.1 using Hansen's bearing capacity equations.

Solution:

Hansen's method extends Meyerhof's solutions by considering the effects of sloping ground surface and tilted base (Figure 10.2) as well as modification of $N_\gamma$ and other factors. For a rectangular footing of $L$ by $B$ ($L > B$) and inclined ground surface, base and load:

$$q_u = c' N_c s_c i_c d_c b_c g_c + \gamma D N_q s_q i_q d_q b_q g_q + 0.5 B \gamma N_\gamma s_\gamma i_\gamma d_\gamma b_\gamma g_\gamma \qquad (10.15)$$

For horizontal ground surface and base $b_c = b_q = b_\gamma = g_c = g_q = g_\gamma = 1$, the general bearing capacity equation becomes the same as Equation 10.7. For the undrained conditions:

$$q_u = 5.14 c_u \left(1 + s_c + d_c - i_c - b_c - g_c\right) + \gamma D \qquad (10.16)$$

The bearing capacity factors of $N_q$ and $N_c$ are the same as Meyerhof above; $N_\gamma$ is:

$$N_\gamma = 1.5(N_q - 1) \tan \phi' \qquad (10.17)$$

The shape and inclination factors are according:

$$s_{c,B} = 1 + \frac{N_q}{N_c} \frac{B}{L} i_{c,B}, \ s_{q,B} = 1 + \frac{B}{L} i_{q,B} \sin \phi', \ s_{\gamma,B} = 1 - 0.4 \frac{B}{L} i_{\gamma,B} \geq 0.6 \qquad (10.18)$$

$$s_{c,L} = 1 + \frac{N_q}{N_c} \frac{L}{B} i_{c,L}, \ s_{q,L} = 1 + \frac{L}{B} i_{q,L} \sin \phi', \ s_{\gamma,L} = 1 - 0.4 \frac{L}{B} i_{\gamma,L} \geq 0.6 \qquad (10.19)$$

For $c_u$, $\phi_u = 0$ soil:

$$s_{c,B} = 0.2\frac{B}{L}i_{c,B}, \quad s_{c,L} = 0.2\frac{L}{B}i_{c,L} \tag{10.20}$$

$$i_{c,i} = i_{q,i} - \frac{1-i_{q,i}}{N_q - 1}, \quad i_{q,i} = \left(1 - \frac{0.5H_i}{V + Ac_b \cot\phi'}\right)^{\alpha_1},$$

$$i_{\gamma,i} = \left(1 - \frac{0.7H_i}{V + Ac_b \cot\phi'}\right)^{\alpha_2} \tag{10.21}$$

where $i$ (in Equations 10.21) $= B$ or $L$, $2 \le \alpha_1 \le 5$, $2 \le \alpha_2 \le 5$, $A$ is the area of the footing base and $c_b$ is the cohesion mobilized in the footing-soil contact area. For the tilted base:

$$i_{\gamma,i} = \left[1 - \frac{(0.7 - \eta°/450°)H_i}{V + Ac_b \cot\phi'}\right]^{\alpha_2} \tag{10.22}$$

For $c_u$, $\phi_u = 0$ soil:

$$i_{c,i} = 0.5 - \sqrt{1 - \frac{H_i}{Ac_b}} \quad i = B \text{ or } L \tag{10.23}$$

In the above equations $B$ and $L$ may be replaced by their effective values expressed by Equations 10.14. The depth factors are specified in two sets.

For $\dfrac{D}{B} \le 1$, $\dfrac{D}{L} \le 1$:

$$d_{c,B} = 1 + 0.4\frac{D}{B}, \quad d_{q,B} = 1 + 2\tan\phi'(1 - \sin\phi')^2\frac{D}{B} \tag{10.24}$$

$$d_{c,L} = 1 + 0.4\frac{D}{L}, \quad d_{q,L} = 1 + 2\tan\phi'(1 - \sin\phi')^2\frac{D}{L} \tag{10.25}$$

For $\dfrac{D}{B} > 1$, $\dfrac{D}{L} > 1$:

$$d_{c,B} = 1 + 0.4\tan^{-1}\frac{D}{B}, \quad d_{q,B} = 1 + 2\tan\phi'(1 - \sin\phi')^2\tan^{-1}\frac{D}{B} \tag{10.26}$$

$$d_{c,L} = 1 + 0.4\tan^{-1}\frac{D}{L}, \quad d_{q,L} = 1 + 2\tan\phi'(1 - \sin\phi')^2\tan^{-1}\frac{D}{L} \tag{10.27}$$

For both sets $d_\gamma = 1.0 \tag{10.28}$

For $c_u$, $\phi_u = 0$ soil:

$$d_{c,B} = 0.4\frac{D}{B}, \quad d_{c,L} = 0.4\frac{D}{L} \tag{10.29}$$

For the sloping ground and tilted base the ground factor $g_i$ and base factor $b_i$ are proposed by Equations 10.30 to 10.33. The angles $\beta$ and $\eta$ are at the same plane either parallel to $B$ or $L$:

$$g_c = 1 - \frac{\beta°}{147°}, \quad g_q = g_\gamma = (1 - 0.5\tan\beta)^5 \tag{10.30}$$

For $c_u$, $\phi_u = 0$ soil:

$$g_c = \frac{\beta°}{147°} \tag{10.31}$$

$$b_c = 1 - \frac{\eta°}{147°}, \ b_q = e^{-2\eta \tan \phi'}, \ b_\gamma = e^{-2.7\eta \tan \phi'} \tag{10.32}$$

For $c_u$, $\phi_u = 0$ soil:

$$b_c = \frac{\eta°}{147°} \tag{10.33}$$

Find $N_\gamma$ from Equation 10.17:

$$N_\gamma = 1.5(N_q - 1)\tan\phi' = 1.5(64.19 - 1.0)\tan 40.0° = 79.53.$$

The inclination factors $i_q = i_\gamma = 1.0$.

Calculate shape and depth factors from Equations 10.18, 10.26 and 10.28:

$$s_q = 1 + \frac{B}{L}i_q \sin\phi' = 1 + \frac{1.0}{1.0}\times 1.0 \times \sin 40.0° = 1.643,$$

$$s_\gamma = 1 - 0.4\frac{B}{L}i_\gamma = 1 - 0.4\frac{1.0}{1.0}\times 1.0 = 0.600.$$

$$d_q = 1 + 2\tan\phi'(1-\sin\phi')^2 \tan^{-1}\frac{D}{B} = 1 + 2\tan 40.0°(1-\sin 40.0°)^2 \tan^{-1}\frac{1.5}{1.0} = 1.210,$$

$$d_\gamma = 1.0.$$

Thus from Equation 10.15 and with $c' = 0$ and $b_c = b_q = b_\gamma = g_c = g_q = g_\gamma = 1$:

(a) $q_u = 16.7 \times 1.5 \times 64.19 \times 1.643 \times 1.210 +$
$0.5 \times 1.0 \times 16.7 \times 79.53 \times 0.600 \times 1.0 = 3595$ kPa.

(b) $q_u = 10.19 \times 1.5 \times 64.19 \times 1.643 \times 1.210 +$
$+0.5 \times 1.0 \times 10.19 \times 79.53 \times 0.600 \times 1.0 = 2194$ kPa.


Problem 10.4

Re-work Problem 10.1 using Vesić's bearing capacity equations.

Solution:

Vesić's bearing capacity solution is similar to Hansen's with minor modifications in $N_\gamma$ and some selected factors. This method seems easier to use, since there is no interrelationships between different factors. In this method Equations 10.15 and 10.16 are used with the following modifications:

$$N_\gamma = 2(N_q + 1)\tan\phi' \tag{10.34}$$

For $c_u$, $\phi_u = 0$ soil with the sloping ground (Figure 10.2):

$$N_\gamma = -2\sin\beta \tag{10.35}$$

The other terms are as follows:

$$s_{c,B} = 1 + \frac{N_q}{N_c}\frac{B}{L}, s_{q,B} = 1 + \frac{B}{L}\tan\phi', s_{\gamma,B} = 1 - 0.4\frac{B}{L} \ge 0.6 \tag{10.36}$$

$$s_{c,L} = 1 + \frac{N_q}{N_c}\frac{L}{B}, s_{q,L} = 1 + \frac{L}{B}\tan\phi', s_{\gamma,L} = 1 - 0.4\frac{L}{B} \ge 0.6 \tag{10.37}$$

For $c_u$, $\phi_u = 0$ soil $s_{c,B} = 0.2\frac{B}{L}$, $s_{c,L} = 0.2\frac{L}{B}$ (10.38)

For $i_{c,i}$ use Hansen's equation; other terms are defined by:

$$i_{q,i} = \left(1 - \frac{H_i}{V + Ac_b\cot\phi'}\right)^{m_i}, i_{\gamma,i} = \left(1 - \frac{H_i}{V + Ac_b\cot\phi'}\right)^{m_i+1} \quad i = B \text{ or } L \tag{10.39}$$

For $c_u$, $\phi_u = 0$ soil $i_{c,i} = 1 - \frac{m_i H_i}{Ac_b N_c}$ $i = B$ or $L$ (10.40)

$$m_B = \frac{2+B/L}{1+B/L}, m_L = \frac{2+L/B}{1+L/B} \tag{10.41}$$

For the sloping ground and tilted base (Figure 10.2):

$$g_c = i_q - \frac{1-i_q}{5.14\tan\phi'}, g_q = g_\gamma = (1 - \tan\beta)^2 \tag{10.42}$$

For $c_u$, $\phi_u = 0$ soil $g_c = \frac{\beta}{5.14}$ (10.43)

$$b_c = 1 - \frac{2\beta}{5.14\tan\phi'}, b_q = b_\gamma = (1 - \eta\tan\phi')^2 \tag{10.44}$$

For $c_u$, $\phi_u = 0$ soil $b_c = \frac{\beta}{5.14}$ (10.45)

The depth factors are the same as Hansen's method. For a circular base, in both Hansen's and Vesić's methods, the dimensions of an equivalent square may be used.

$N_\gamma = 2(N_q + 1)\tan\phi' = 2(64.19 + 1.0)\tan 40.0° = 109.40.$

Shape factors are calculated from Equations 10.36:

$s_q = 1 + (B/L)\tan\phi' = 1 + (1.0/1.0)\tan 40.0° = 1.839$,

$s_\gamma = 1 - 0.4(B/L) = 1 - 0.4(1.0/1.0) = 0.60.$

Depth factors are the same as Hansen's method: $d_q = 1.210$, $d_\gamma = 1.0$.

(a) $q_u = 16.7 \times 1.5 \times 64.19 \times 1.839 \times 1.210 + 0.5 \times 1.0 \times 16.7 \times 109.4 \times 0.6 \times 1.0 = 4126 \text{ kPa}.$

(b) $q_u = 10.19 \times 1.5 \times 64.19 \times 1.839 \times 1.210 + 0.5 \times 1.0 \times 10.19 \times 109.4 \times 0.6 \times 1.0 = 2518 \text{ kPa}.$

Problem 10.5

Using Hansen's method calculate the base capacity of a square pile of 0.4 m width and 10 m length for the following two cases:
(a) the water table is well below the pile base with $\gamma = 16.7 \text{ kN/m}^3$,

(b) the water table is at the ground surface with $\gamma_{sat} = 20$ kN/m$^3$. Ignore the $N_\gamma$ term. $c' = 0$, $\phi' = 40°$.

Solution:

For $c'$, $\phi'$ soil the general form of the ultimate bearing capacity of the pile base is:

$$q_b = c'N_c d_c + \eta p'_o N_q d_q \tag{10.46}$$

where $p'_o$ is the effective vertical stress at the base level due to the weight of the soil and $\eta$ represents the effect of the at-rest lateral earth pressure. The $N_\gamma$ term is negligible in comparison with the other terms. The Hansen's bearing capacity equations may be used where the pile length $L$ replaces $D$ in the depth factors and $\eta = 1$.

For $c_u$, $\phi_u = 0$ soil (undrained conditions) the ultimate bearing capacity of the pile base is reduced to:

$$q_b = c_u N_c d_c \tag{10.47}$$

The term $N_c d_c$ may be assumed 9 for most practical purposes.

$A_b = 0.4 \times 0.4 = 0.16\,\text{m}^2$.

Using Equations 10.9 we find $N_q = 64.19$.

Depth factors from Equations 10.26:

$d_q = 1 + 2\tan 40.0°(1 - \sin 40.0°)^2 \tan^{-1}(10.0/0.40) = 1.328$.

(a) $p'_o = 16.7 \times 10.0 = 167.0$ kPa. Thus using Equation 10.46:

$q_b = 167.0 \times 64.19 \times 1.328 = 14236$ kPa.

Base capacity:

$P_b = q_b A_b = 14236 \times 0.16 = 2278$ kN.

(b) $p'_o = 10.19 \times 10.0 = 101.9$ kPa.

$q_b = 101.9 \times 64.19 \times 1.328 = 8686$ kPa.

Base capacity:

$P_b = q_b A_b = 8686 \times 0.16 = 1390$ kN.

**Problem 10.6**

Re-work Problem 10.5 using Vesić's method with $I_{rr} = 20$.

Solution:

The values of $N'_c$ and $N'_q$ (instead of $N_c$ and $N_q$ in Equation 10.46) derived by Vesić are:

$$N'_c = (N'_q - 1)\cot\phi' \tag{10.48}$$

$$N'_q = \frac{3}{3 - \sin\phi'}\left\{\exp[(\pi/2 - \phi')\tan\phi']\tan^2(45° + \phi'/2)I_{rr}^{\frac{4\sin\phi'}{3(1+\sin\phi')}}\right\} \tag{10.49}$$

The term $I_{rr}$ is called the reduced rigidity index and is defined by:

$$I_{rr} = \frac{I_r}{1 + \varepsilon_V I_r} \tag{10.50}$$

where $\varepsilon_V$ is the volumetric strain of the soil at the vicinity of the pile base at failure and $I_r$ is a rigidity index according:

$$I_r = \frac{G}{c' + p'_o \tan \phi'}$$

(10.51)

The term $G$ represents the shear modulus of the soil. Note that in undrained conditions $\varepsilon_V = 0$. The parameter $\eta$ is a function of the coefficient of soil pressure at-rest condition $k_o$ in the following form:

$$\eta = \frac{1 + 2k_o}{3}$$

(10.52)

The Vesić's $N'_c$ value for undrained conditions is:

$$N'_c = \frac{4(\ln I_{rr} + 1)}{3} + \frac{(\pi + 2)}{2}$$

(10.53)

From Equation 10.49 we find:

$$N'_q = \frac{3}{3 - \sin 40.0^\circ} \left\{ \exp\left[ \left( \pi/2 - \frac{\pi \times 40.0^\circ}{180.0^\circ} \right) \tan 40.0^\circ \right] \tan^2 65.0^\circ \times 20.0^{\frac{4 \sin 40.0^\circ}{3(1 + \sin 40.0^\circ)}} \right\},$$

$N'_q = 58.10$.

The lateral soil pressure coefficient at-rest $k_o$ may be calculated from:

$$k_o = 1 - \sin \phi'$$

(10.54)

Thus $k_o = 1 - \sin 40.0^\circ = 0.357$.

From Equation 10.52:

$\eta = (1 + 2 \times 0.357)/3 = 0.571$.

(a) $q_b = 0.571 \times 167.0 \times 58.1 \times 1.328 = 7357$ kPa.

$P_b = 7357 \times 0.16 = 1177$ kN.

(b) $q_b = 0.571 \times 101.9 \times 58.1 \times 1.328 = 4489$ kPa.

$P_b = 4489 \times 0.16 = 718$ kN.

Problem 10.7

A pile of length 15 m is embedded in a sand layer with the following properties:

Density index $I_D = 0.75$, critical friction angle $\phi'_{cr} = 33^\circ$, $\gamma = 17$ kN/m$^3$, $\gamma_{sat} = 20$ kN/m$^3$.

Calculate the base bearing capacity if:

(a) there is no water in the vicinity of the pile base,

(b) the water table is at the ground surface.

Use the iteration based method of Fleming et al. (1992).

Solution:

The critical friction angle $\phi'_{cr}$ is a constant property of the soil being independent from the initial void ratio and stress level. The magnitude of this parameter can be found from:

$$\sin \phi'_{cr} = \frac{3M}{6+M},$$

where $M = q'/p'$ (Equation 4.12), $q'$ equals to $(\sigma'_1 - \sigma'_3)$ and $p'$ is the effective mean stress $(\sigma'_1 + 2\sigma'_3)/3$.

The appropriate value of internal friction angle $\phi'$ under the pile base is estimated from the following equation:

$$\phi' = \phi'_{cr} + 3I_D\left[5.4 - \ln(p'/p_a)\right] - 3 \quad \text{(degrees)} \tag{10.55}$$

where $p_a$ is the atmospheric pressure ($\approx 100$ kPa). The base bearing capacity is defined by:

$$q_b = p'_o N_q \tag{10.56}$$

in which $N_q$ is the bearing capacity factor corresponding to the internal friction angle $\phi'$ mobilized beneath the base. The effective mean stress $p'$ in the vicinity of the pile base is taken as the geometric mean of the base bearing capacity and effective overburden pressure $p'_o$:

$$p' = \sqrt{N_q\, p'_o} \tag{10.57}$$

To evaluate $N_q$, an iterative process is carried out by first assuming an initial value for $N_q$ and calculating $p'$ from Equation 10.57. The corresponding $\phi'$ is then calculated from Equation 10.55 and, using an appropriate $\phi'\text{-}N_q$ relationship, the value of $N_q$ can be obtained. If this value is not sufficiently close to the assumed $N_q$, iteration is continued until the difference in $N_q$ between successive cycles becomes insignificant. The $\phi'\text{-}N_q$ relationship used is taken from the theory of ultimate bearing capacity developed by Berezantzev et al. (1961) illustrated in Figure 10.3.

(a) $p'_o = 17.0 \times 15.0 = 255.0$ kPa. Assume $N_q = 79.0$, thus

$$p' = \sqrt{N_q\, p'_o} = \sqrt{79.0 \times 255.0} = 2266.5 \text{ kPa.}$$

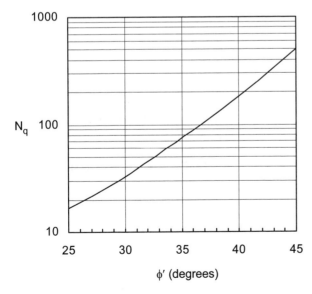

Figure 10.3. Bearing capacity factor $N_q$ used for piles in sand (Berezantzev et al., 1961).

$\phi' = 33.0° + 3 \times 0.75[5.4 - \ln(2266.5/100.0)] - 3.0° = 35.1°.$

From Figure 10.3 $N_q = 75.0$.

For the second iteration we assume $N_q = 75.0$, therefore:

$p' = \sqrt{N_q p'_o} = \sqrt{75.0} \times 255.0 = 2208.4$ kPa.

$\phi' = 33.0° + 3 \times 0.75[5.4 - \ln(2208.4/100.0)] - 3.0° = 35.2°.$

Select $N_q = 75.0$.

$q_b = p'_o N_q = 255.0 \times 75.0 = 19125$ kPa $= 19.1$ MPa.

(b) $p'_o = (20.0 - 9.81) \times 15.0 = 152.8$ kPa. Assume $N_q = 100.0$, thus

$p' = \sqrt{N_q p'_o} = \sqrt{100.0} \times 152.8 = 1528.0$ kPa.

$\phi' = 33.0° + 3 \times 0.75[5.4 - \ln(1528.0/100.0)] - 3.0° = 36.0°.$

From Figure 10.3 $N_q < 100.0$.

For the second iteration we assume $N_q = 90.0$, therefore:

$p' = \sqrt{N_q p'_o} = \sqrt{90.0} \times 152.8 = 1449.6$ kPa.

$\phi' = 33.0° + 3 \times 0.75[5.4 - \ln(1449.6/100.0)] - 3.0° = 36.1°.$

Select $N_q = 88.0$.

$q_b = p'_o N_q = 152.8 \times 88.0 = 13446$ kPa $= 13.4$ MPa.

## Problem 10.8

Estimate the ultimate pile capacity of a 30 m concrete pile with 0.4 m diameter in an offshore structure where the submerged unit weight is 8.3 kN/m$^3$. The profile of the undrained shear strength, which changes linearly between the measured points, is:

| Depth (m) | 0 | 6 | 18 | 24 | 30 |
|-----------|-----|-----|-----|-----|-----|
| $c_u$ (kPa) | 200 | 440 | 440 | 220 | 220 |

Solution:

In the conventional analysis called $\alpha$ method the average limiting shear stress $\tau_s$ mobilized on the shaft is estimated empirically as a fraction of the undrained cohesion $c_u$:

$$\tau_s = \alpha c_u \qquad\qquad (10.58)$$

This method has been improved (Fleming et al., 1992) relating $\alpha$ to the strength ratio defined by $c_u / p'_o$ according:

$$\alpha = \left(\frac{c_u}{p'_o}\right)_{NC}^{0.5} \left(\frac{c_u}{p'_o}\right)^{-0.5} \qquad c_u / p'_o \leq 1 \qquad (10.59)$$

$$\alpha = \left(\frac{c_u}{p'_o}\right)_{NC}^{0.5} \left(\frac{c_u}{p'_o}\right)^{-0.25} \qquad c_u / p'_o > 1 \qquad (10.60)$$

where the subscript $NC$ represents the normally consolidated state. For normally consolidated soil $\alpha = 1$. The shaft capacity may be expressed by:

$$P_s = \sum_{i=1}^{i=n} (\alpha c_u A_s)_i \tag{10.61}$$

where $\alpha_i$, $c_{ui}$ are the average values related to a finite length of a pile in specified depth, and $A_{si}$ is the perimeter surface area of the finite length. Based on the experimental data for driven piles reported by Randolph & Murphy (1985) Equations 10.59 and 10.60 may be simplified to the following:

$$\alpha = 0.5 \left( \frac{c_u}{p'_o} \right)^{-0.5} \qquad \frac{c_u}{p'_o} \leq 1 \tag{10.62}$$

$$\alpha = 0.5 \left( \frac{c_u}{p'_o} \right)^{-0.25} \qquad \frac{c_u}{p'_o} > 1 \tag{10.63}$$

The results of computations are tabulated where the total shaft capacity is 4824 kN.

| Depth (m) | $c_u$ (kPa) | $p'_o$ (kPa) | $c_u / p'_o$ | $\alpha$ | $\tau_s$ (kPa) | $A_s$ (m$^2$) | $\tau_s \times A_s$ (kN) |
|---|---|---|---|---|---|---|---|
| 0-6 | 320 | 24.9 | 12.85 | 0.264 | 84.5 | 7.54 | 637.1 |
| 6-18 | 440 | 99.6 | 4.42 | 0.345 | 151.8 | 15.08 | 2289.1 |
| 18-24 | 330 | 174.3 | 1.89 | 0.426 | 140.6 | 7.54 | 1060.1 |
| 24-30 | 220 | 224.1 | 0.98 | 0.505 | 111.1 | 7.54 | 837.7 |
| | | | | | | | Total $\approx$ 4824 |

Sample calculation for depth 18 m - 24 m:

$c_u$ (average) $= (440.0 + 220.0)/2 = 330.0$ kPa.

$p'_o = 8.3 \times (18.0 + 24.0)/2 = 174.3$ kPa.

$c_u / p'_o = 330.0/174.3 = 1.89$.

From Equation 10.63:

$\alpha = 0.5 \times (1.89)^{-0.25} = 0.426$.

$\tau_s = 0.426 \times 330.0 = 140.6$ kPa.

$A_s = (\pi \times 0.4)(24.0 - 18.0) = 7.54$ m$^2$.

$\tau_s \times A_s = 140.6 \times 7.54 = 1060.1$ kN.

The ultimate bearing capacity of the pile base is calculated from Equation 10.47:

$q_b = c_u N_c d_c = c_u \times 9 = 220.0 \times 9 = 1980$ kPa.

$P_b = 1980.0 \times \pi 0.4^2 / 4 = 248.8$ kN.

The ultimate bearing capacity $P_u$ is the sum of the shaft capacity $P_s$ and the base capacity $P_b$:

$P_u = P_b + P_s = 248.8 + 4824 \approx 5073$ kN $= 5.07$ MN.

Problem 10.9

It is required to estimate the length of a frictional concrete pile of 0.4 m diameter embedded in cohesionless soil. The ultimate bearing capacity of the pile is 1850 kN. The soil is comprised of 10 m fine sand with $\gamma = 16.5$ kN/m$^3$ and $\phi' = 30°$ underlain by a course sand of $\gamma = 18.8$ kN/m$^3$ and $\phi' = 36°$.

Assume $K$ (lateral earth pressure coefficient) = 1.5 for both layers.

Solution:

An alternative effective stress analysis suggested by Burland (1973) assumes no effective cohesion on the pile shaft due to the remoulding effects of pile installation. The ultimate frictional shear stress mobilized at a specific depth on the shaft is:

$$\tau_s = \sigma'_h \tan\delta' = Kp'_o \tan\delta' = \beta p'_o \tag{10.64}$$

where $\sigma'_h$ is the effective normal stress horizontally applied by the soil on the pile, $\delta'$ is the effective friction angle mobilized on the pile surface, $p'_o$ is the effective vertical stress, and $K$ represents the lateral earth pressure coefficient. For bored piles in heavily consolidated soils a value of $K = (1 + k_o)/2$ is recommended. For driven piles experimental results give a range for $K$ from 1.5 to 1.9.

Find the average effective vertical stress at the fine sand layer:

$p'_o = 16.5 \times (10.0/2) = 82.5$ kPa.

The average mobilized shear stress on the shaft from Equation 10.64 and by assuming $\delta' = \phi' = 30.0°$ is:

$\tau_s = Kp'_o \tan\delta' = 1.5 \times 82.5 \times \tan 30.0° = 71.4$ kPa. The corresponding shaft capacity is:

$P_{s1} = \pi \times 0.4 \times 10.0 \times 71.4 = 897.2$ kN $< 1850.0$ kN.

The pile must be extended to the coarse sand to the depth of $l$ with shaft capacity of:

$P_{s2} = 1850.0 - 897.2 = 952.8$ kN.

$p'_o = 16.5 \times 10.0 + 18.8 \times l/2 = 165.0 + 9.4l$ kPa.

$\tau_s = Kp'_o \tan\delta' = 1.5(165.0 + 9.4l) \tan 36.0°$ kPa.

$P_{s2} = 952.8 = \pi \times 0.4 \times l \times 1.5(165.0 + 9.4l) \tan 36.0°$,

$l^2 + 17.55l - 74.01 = 0 \rightarrow l = 3.5$ m.

Thus the total length of the pile is $10.0 + 3.5 = 13.5$ m.

Problem 10.10

A pile group of Figure 10.4 carries 600 kN vertical force at $x = y = 0.7$ m. The piles are of equal diameter with $s_x = 1.4$ m and $s_y = 1.2$ m. Calculate the vertical load at each pile.

Solution:

For the load distribution through a rigid pile cap Equation 5.50 may be used in the following form:

$$P_i = S_i \left( \frac{M_y}{I_y} x_i - \frac{M_x}{I_x} y_i + \frac{P}{S} \right) \tag{10.65}$$

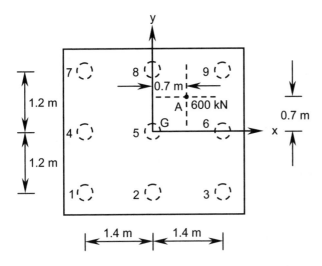

Figure 10.4. Problem 10.10.

where $P$ is the total vertical load, $M_x$ and $M_y$ are the moments about the $x$ and $y$ axes (Figure 10.4), $x_i$, $y_i$ are the coordinates of the centriod of each pile, $I_x$ and $I_y$ are the second moments of area of the pile group about the $x$ and $y$ axes, $S_i$ is the cross-sectional area of each pile, $S$ is the total cross-sectional area of the pile group (excludes the area of the pile cap) and $P_i$ is the vertical load taken by the pile $i$. The origin of the coordinate system is at the centroid of the pile group, which may be different from the centriod of the rigid cap. For $n$ piles of equal cross-sectional area:

$$P_i = \frac{M_y}{\sum x_i^2} x_i - \frac{M_x}{\sum y_i^2} y_i + \frac{P}{n} \tag{10.66}$$

$\left|M_x\right| = \left|M_y\right| = 600.0 \times 0.7 = 420.0$ kN.m.

Considering the right-hand rule sign convention:

$M_x = -420.0$ kN.m, $M_y = 420.0$ kN.m.

$\sum x_i^2 = 6 \times 1.4^2 = 11.76 \, \text{m}^2$,

$\sum y_i^2 = 6 \times 1.2^2 = 8.64 \, \text{m}^2$.

From Equation 10.66:

$$P_1 = \frac{420.0}{11.76} \times -1.4 - \frac{-420.0}{8.64} \times -1.2 + \frac{600.0}{9} = -41.7 \, \text{kN}.$$

$$P_2 = \frac{420.0}{11.76} \times 0.0 - \frac{-420.0}{8.64} \times -1.2 + \frac{600.0}{9} = 8.3 \, \text{kN}.$$

$$P_3 = \frac{420.0}{11.76} \times 1.4 - \frac{-420.0}{8.64} \times -1.2 + \frac{600.0}{9} = 58.3 \, \text{kN}.$$

Similarly,

$P_4 = 16.7$ kN, $P_5 = 66.7$ kN,

$P_6 = 116.7$ kN, $P_7 = 75.0$ kN,

$P_8 = 125.0$ kN, and $P_9 = 175.0$ kN.

Problem 10.11

A pile group has 3 coaxial piles all having equal length of 10 m and are spaced equally at 1.5 m. The central pile carries 600 kN; the side piles each has 450 kN. Calculate the settlement of the central pile assuming skin resistance load mechanism.
$\mu$ (Poisson's ratio) = 0.3, $E_s = 8$ MPa.

Solution:

The base settlement may be calculated using the Mindlin stress distribution. After dividing the soil under the base of the pile into a finite number of layers the vertical stress is computed using Table 10.1 (or relevant equations, see Aysen (2002) regarding tables for different values of $\mu$ and influence factors for end bearing piles).

Table 10.1. Vertical stress influence factors due to uniform shear force applied in the interior of the soil ($\mu = 0.3$).

| $n \rightarrow 0.00$ $m \downarrow$ | 0.02 | 0.04 | 0.06 | 0.08 | 0.10 | 0.15 | 0.20 | 0.50 | 1.00 | 2.00 |
|---|---|---|---|---|---|---|---|---|---|---|
| 1.0 $\infty$ | 6.8419 | 3.4044 | 2.2673 | 1.6983 | 1.3567 | 0.8998 | 0.6695 | 0.2346 | 0.0686 | 0.0076 |
| 1.1 1.9219 | 1.8611 | 1.7072 | 1.5134 | 1.3211 | 1.1503 | 0.8368 | 0.6419 | 0.2335 | 0.0728 | 0.0091 |
| 1.2 0.9699 | 0.9403 | 0.9166 | 0.8825 | 0.8400 | 0.7922 | 0.6688 | 0.5588 | 0.2292 | 0.0760 | 0.0105 |
| 1.3 0.6430 | 0.6188 | 0.6099 | 0.5992 | 0.5850 | 0.5675 | 0.5157 | 0.4597 | 0.2207 | 0.0782 | 0.0120 |
| 1.4 0.4867 | 0.4558 | 0.4507 | 0.4461 | 0.4396 | 0.4316 | 0.4063 | 0.3761 | 0.2082 | 0.0796 | 0.0134 |
| 1.5 0.3766 | 0.3561 | 0.3533 | 0.3510 | 0.3476 | 0.3432 | 0.3291 | 0.3115 | 0.1834 | 0.0800 | 0.0148 |
| 1.6 0.3339 | 0.2895 | 0.2878 | 0.2863 | 0.2843 | 0.2817 | 0.2732 | 0.2621 | 0.1777 | 0.0796 | 0.0160 |
| 1.7 0.2664 | 0.2438 | 0.2414 | 0.2399 | 0.2384 | 0.2369 | 0.2313 | 0.2239 | 0.1623 | 0.0784 | 0.0172 |
| 1.8 0.2025 | 0.2065 | 0.2054 | 0.2044 | 0.2038 | 0.2026 | 0.1989 | 0.1938 | 0.1479 | 0.0766 | 0.0182 |
| 1.9 0.1847 | 0.1794 | 0.1785 | 0.1777 | 0.1768 | 0.1760 | 0.1733 | 0.1696 | 0.1347 | 0.0744 | 0.0191 |
| 2.0 0.1634 | 0.1565 | 0.1561 | 0.1556 | 0.1551 | 0.1545 | 0.1525 | 0.1498 | 0.1229 | 0.0718 | 0.0199 |

The vertical stress component at any point with a horizontal distance $r$ from the load $Q$ and a depth $z > L$ (pile length) is:

$$\sigma'_z = \frac{Q}{L^2} I_q \tag{10.67}$$

where $I_q$ is given by the above table. The coordinates of the point of interest are expressed in the dimensionless parameters of $m = z/L$, and $n = r/L$. The settlement of the base is approximately:

$$w_b = \sum_{i=1}^{i=n} (\frac{\sigma'_z}{E_s} l)_i \tag{10.68}$$

where $l$ is the thickness of the finite layer, $\sigma'_z$ is the increase in the effective vertical stress at the mid-point of the layer due to loading, and $E_s$ is the Modulus of Elasticity of each finite layer.

For a uniform $E_s$ and assuming: $Q$ = allowable load applied to the pile = $P_a$, substituting $\sigma'_z$ from Equation 10.67 into Equation 10.68:

$$w_b = \frac{P_a}{L^2 E_s} \sum_{i=1}^{i=n} (II_q)_i .$$

If $l_1 = l_2 = l_3 = \cdots = l_n$, and

$l_1 + l_2 + l_3 + \cdots + l_n = L$ (length of the pile):

$$w_b = \frac{P_a}{LE_s} \sum_{i=1}^{i=n} \left( \frac{I_q}{n} \right)_i = \frac{P_a}{LE_s} I_{q(average)} \tag{10.69}$$

which simplifies the use of the above or similar tables.

The soil beneath the base is divided into 5 layers of equal thickness of:

$l_1 = l_2 = l_3 = l_4 = l_5 = L/5 = 2.0$ m.

$n_1$ (due to the central pile) = $r/L = 0.0/10.0 = 0$.

$n_2$ and $n_3$ (due to side piles) = $r/L = 1.5/10.0 = 0.15$.

$m_1$ (centre of layer 1) = $z/L = 11.0/10.0 = 1.1$,

$m_2 = 1.3$, $m_3 = 1.5$, $m_4 = 1.7$, and $m_5 = 1.9$.

The results are tabulated below:

| Layer | $z$ (m) | $m = z/L$ | $I_q$ $(n = 0.0)$ | $I_q$ $(n = 0.15)$ |
|---|---|---|---|---|
| 1 | 11.0 | 1.1 | 1.9219 | 0.8368 |
| 2 | 13.0 | 1.3 | 0.6430 | 0.5157 |
| 3 | 15.0 | 1.5 | 0.3766 | 0.3291 |
| 4 | 17.0 | 1.7 | 0.2664 | 0.2313 |
| 5 | 19.0 | 1.9 | 0.1847 | 0.1733 |
| | | Total: | 3.3926 | 2.0862 |
| | | Average: | 0.6785 | 0.4172 |

Using Equation 10.69:

$$w_b \text{ (due to middle pile)} = \frac{600.0}{10.0 \times 8000.0} \times 0.6785 = 5.1 \times 10^{-3} \text{ m} = 5.1 \text{ mm}.$$

$$w_b \text{ (due to side piles)} = 2 \times \frac{450.0}{10.0 \times 8000.0} \times 0.4172 = 4.7 \times 10^{-3} \text{ m} = 4.7 \text{ mm}.$$

The base settlement of the central pile is:

$w_b \text{ (total)} = 5.1 + 4.7 = 9.8 \approx 10$ mm.

## 10.3 REFERENCES AND RECOMMENDED READINGS

Anagnostpoulos, C. & Georgiadis, M. 1993. Interaction of axial and lateral pile responses. *Journal SMFED, ASCE*, 119(4): 793-798.

API, 1984. *API recommended practice for planning, designing and constructing fixed offshore platforms*. 15th edition. API RP2A page 115. American Petroleum Institute.

Aysen, A. 2002. *Soil mechanics: Basic concepts and engineering applications*. Lisse: Balkema.

Berezantzev, V.C., Khristoforov, V. & Golubkov, V. 1961. Load bearing capacity and deformation of piled foundations. *Proc. 5th intern. conf. SMFE*, 2: 11-15.

Bergdahl, U. & Hult, G. 1981. Load tests on friction piles in clay. *Proc. 10th intern. conf. SMFE*, 2: 625-630. Stockholm.

Bolton, M.D. 1986. The strength and dilatancy of sands. *Geotechnique*, 36(1): 65-78.

Bowles, J.E. 1996. *Foundation analysis and design*. 5th edition. New York: McGraw-Hill.

Burland, J.B. 1973. Shaft friction piles in clay-a simple fundamental approach. *Ground engineering*, 6(3): 30-42.

Burland, J.B. & Burbidge, M.C. 1985. Settlements of foundations on sands and gravel. *Proc.*, Part 1, 78: 1325-1381. London: Institution of Civil Engineers.

Butterfield, R. & Douglas, R.A. 1981. Flexibility coefficients for the design of piles and pile groups. *CIRIA technical note*, 108.

Cooke, R.W. 1974. Settlement of friction pile foundations. *Proc. conf. on tall buildings*: 7-19. Kuala Lumpur.

D'Appolonia, E. & Romualdi, J.P. 1963. Load transfer in end bearing steel H piles. *Journal SMFED, ASCE*, 89(SM2): 1-26.

DeBeer, E.E. & Martens, A. 1957. Method of computation of an upper limit for the influence of heterogeneity of sand layers on the settlement of bridges. *Proc. 4th intern. conf. SMFE*, 1:275-281. London: Butterworths.

DeNicola, D.A. & Randolph, M.F. 1993. Tensile and compression shaft capacity of piles in sand. *Journal SMFED, ASCE*, 119(12): 1952-1973.

Fleming, W.G.K. & Thorburn, S. 1983. Recent piling advances: state of the art report. *Proc. conf. on advances in piling and ground treatment for foundations*. London: Institute of Civil Engineers.

Fleming, W.G.K., Weltman, A.G., Randolph M.F. & Elson, W.K. 1992. *Piling Engineering*. 2nd edition. New York: Blackie Academic & Professional.

Francescon, M. 1983. *Model pile tests in clay: stresses and displacements due to installation and axial loading*. Ph.D. Thesis, University of Cambridge.

Ghazavi, M. 1997. *Static and dynamic analysis of piled foundations*. Ph.D. Thesis, University of Queensland, Australia.

Gibbs, H.J. & Holtz, W.G. 1957. Reseach on determining the density of sands by spoon penetration testing. *Proc. 4th intern. conf. SMFE*, 1:35-39.

Gupte, A.A. 1984. Construction, design and application of post tensioned auger cast soil anchors. *Intern. symp. of prestressed rock and soil anchors*. Post-Tensioning Institute, Des Plaines, IL, USA.

Gupte, A.A. 1989. Design, construction and applications of auger piling system for earth retention and underpinning of structures. In J. Burland & J. Mitchell (eds), *Piling and deep foundations*, 1: 63-72. Rotterdam: Balkema.

Hansen, J.B. 1961. A general formula for bearing capacity. *Danish geotechnical institute*. Copenhagen, Denmark. Bulletin (11): 38-46.

Hansen, J.B. 1970. A revised and extended formula for bearing capacity. *Danish geotechnical institute*. Copenhagen, Denmark. Bulletin (28): 5-11.

Jumikis, A.R. 1971. *Foundation engineering*. Scranton, Pa: Intext Educational Publishers.

Konrad, J.M. & Roy, M. 1987. Bearing capacity of friction piles in marine clay. *Geotechnique*, 37(2): 163-175.

Lee, S.L., Chow, Y.K., Karunarante, G.P. & Wong, K.Y. 1988. Rational wave equation model for pile driving analysis. *Journal SMFED, ASCE*, 114(3): 306-325.

Liao, S.S. & Whitman, R.V. 1986. Overburden correction factors for sand. *Journal SMFED, ASCE*, 112(3): 373-377.

Lunne, T. & Edie, O. 1976. Correlations between cone resistance and vane shear strength in some Scandinavian soft to medium stiff clays. *Canadian geotechnical journal*, 13(4): 430-441.

MacDonald, D.H. & Skempton, A.W. 1955. A survey of comparisons between calculated and observed settlements of structures on clay. *Conf. on correlation of calculated and observed stresses and displacements*: 318-337. London: Institute of Civil Engineers.

Mansur, C.I. & Hunter, A.H. 1970. Pile tests-Arkansas river project. *Journal SMFD, ASCE*, 96(SM5): 1545-1582.

Meyerhof, G.G. 1951. The ultimate bearing capacity of foundations. *Geotechnique*, 2: 301-332.

Meyerhof, G.G. 1953. The bearing capacity of foundations under eccentric and inclined loads. *Proc. 3$^{rd}$ intern. conf. SMFE*, 1: 440-445. Zurich.

Meyerhof, G.G. 1956. Penetration tests and bearing capacity of cohesionless soils. *Journal SMFED, ASCE*, 82(SM1): 1-19.

Meyerhof, G.G. 1963. Some recent research on the bearing capacity of foundations. *Canadian geotechnical journal*, 1(1): 16-26.

Meyerhof, G.G. 1965. Shallow foundations. *Journal SMFED, ASCE*, 91(SM2): 21-31.

Meyerhof, G.G. 1976. Bearing capacity and settlement of pile foundations. *Journal SMFED, ASCE*, 91(GT3): 195-228.

Mori, H & Inamura, T. 1989. Construction of cast-in-place piles with enlarged bases. In J. Burland & J. Mitchell (eds), *Piling and deep foundations,* 1: 85-92. Rotterdam: Balkema.

Peck, R.B, Hanson, W.E. & Thornburn, T.H. 1974. *Foundation engineering.* New York: John Wiley & Sons.

Powrie, W. 1997. *Soil mechanics-concepts and applications.* London: E & FN Spon.

Ramiah, B.K. & Chickanagappa, L.S. 1982. *Handbook of soil mechanics and foundation engineering.* Rotterdam: Balkema.

Poulos, H.G. 1982. The influence of shaft length on pile load capacity in clays. *Geotechnique*, 32(2): 145-148.

Poulos, H.G. & Davis, E.H. 1980. *Pile foundation analysis and design.* John Wiley & Sons.

Randolph, M.F. & Murphy, B.S. 1985. Shaft capacity of driven piles in clay. *Proc. 17$^{th}$ offshore technology conf.*: 371-378.

Randolph, M.F. & Simons, H.A. 1986. An improved soil model for one-dimensional pile driving analysis. *Proc. 3$^{rd}$ intern. conf. on numerical methods in offshore piling*: 3-15. Nantes, France.

Randolph, M.F. & Wroth, C.P. 1978. Analysis of deformation of vertically loaded piles. *Journal SMFED, ASCE*, 104(GT12): 1465-1488.

Robertson, P.K. & Camanella, R.G. 1983. Interpretation of cone penetration tests. *Canadian geotechnical journal*, 20(4): 718-745.

Robertson, P.K., Campanella, R.G. & Wightman, A. 1983. SPT-CPT correlations. *Journal SMFED, ASCE*, 109(11): 1449-1459.

Schmertmann, J.H. 1970. Static cone to compute static settlement over sand. *Journal SMFED, ASCE*, 96(SM3): 1011-1043.

Schmertmann, J.H. 1975. Measurement of in-situ shear strength. *Proc. conf. on in-situ measurement of soil properties*, 2: 57-138. New York: ASCE.

Schmertmann, J.H., Hartman, J.P. & Brown, P.R. 1978. Improved strain influence factor diagrams. *Journal SMFED, ASCE*, 104(8): 1131-1135.

Seed, H.B., Tokimatsu, K., Harder, L.F. & Chung, R.M. 1984. The influence of SPT procedures in soil liquefaction evaluations. *Journal SMFED, ASCE*, 111(12): 1425-1445.

Skempton. A.W. 1951. The bearing capacity of clays. *Proc. building research congress*, 1: 180-189. London, UK.

Skempton. A.W. 1959. Cast in-situ bored piles in London Clay. *Geotechnique*, 9: 153-173.

Skempton. A.W. 1986. Standard penetration test procedures and the effects in sands of overburden pressure, relative density, particle size, aging and overconsolidation. *Geotechnique*, 36(3): 425-447.

Smith, E.A.L. 1960. Pile driving analysis by the wave equation. *Journal SMFED, ASCE*, 86(SM4): 35-61.

Stroud, M.A. 1989. The standard penetration test: its application and interpretation. *Penetration testing in the UK*: 29-49. London: Thomas Telford.

Terzaghi, K. 1943. *Theoretical soil mechanics*. New York: John Wiley & Sons.

Terzaghi, K. & Peck, R. B. 1967. *Soil mechanics in engineering practice*. 2nd edition. New York: John Wiley & Sons.

Terzaghi, K., Peck, R. B., & Mesri, G. 1996. *Soil mechanics in engineering practice*. 3rd edition. New York: John Wiley & Sons.

Thorburn, S. & McVicar, R.S.L. 1971. Pile load tests to failure in the Clyde alluvium. *Proc. conf. on behaviour of piles*: 1-7, 53-54. London: Institute of Civil Engineers.

Tomilinson, M.J. 1977. *Pile design and construction practice*. 4th edition. London: E. & FN Spon.

Tomilinson, M.J. 1995. *Foundation design and construction*. 6th edition. Harlow, Essex: Longman Scientific & Technical.

Vesić, A.S. 1973. Analysis of ultimate loads of shallow foundations. *Journal SMFED, ASCE*, 99(SM1): 45-73.

Vesić, A.S. 1975. Principles of pile foundation design. *Soil mechanics services*. School of Engineering, Duke University, Durham. (38): page 48.

Walsh, H. 1981. Tolerable settlement of buildings. *Journal SMFED, ASCE*, 107(ET11): 1489-1504.

Weltman, A.J. & Little, J.A.1977. A review of bearing pile types. *DoE/CIRIA piling development group*, report PG1.

Whitaker, T. & Cooke, R.W. 1966. An investigation of the shaft and base resistance of large bored piles in London clay. *Proc. symp. on larged bored piles*: 7-49. London: Institution of Civil Engineers.

# Index